ENGINEERING TOOLS, TECHNIQUES AND TABLES

LIQUID CRYSTAL PHOTONICS

ENGINEERING TOOLS, TECHNIQUES AND TABLES

Additional books in this series can be found on Nova's website under the Series tab.

Additional e-books in this series can be found on Nova's website under the e-book tab.

ENGINEERING TOOLS, TECHNIQUES AND TABLES

LIQUID CRYSTAL PHOTONICS

VLADIMIR G. CHIGRINOV

nova publishers
New York

NOTICE TO THE READER

Additional color graphics may be available in the e-book version of this book.

Library of Congress Cataloging-in-Publication Data

Chigrinov, V. G. (Vladimir G.), author.
Liquid crystal photonics / Vladimir G. Chigrinov (Department of Electronic and Computer Engineering, Hong Kong University of Science and Technology, Clear Water Bay, Kowloon, Hong Kong).
pages cm. -- (Engineering tools, techniques and tables)
Includes bibliographical references and index.
ISBN 978-1-62948-315-3 (hardcover)
1. Liquid crystals. 2. Photonics. I. Title.
TA418.9.L54C45 2014
621.36'5--dc23

2014033257

Published by Nova Science Publishers, Inc. † New York

CONTENTS

PREFACE

Liquid crystal (LC) devices (LCD) are used in almost all electronic devices and information systems. This outstanding success was achieved mainly via an enthusiasm of physicists, chemists and engineers who not only studied unique properties of liquid crystals but also thought about their optimization providing technical privileges of liquid crystal devices. Despite the relative maturity of LCD technologies, however, many novel research is being performed on many aspects of LCDs, in particular in LC photonics applications. It is our desire to add a new book to this collection. The book provides engineers, physicists, and device designers with the most up-to-date descriptions of the dielectric, optical, and viscoelastic properties of LCs, for photonics applications and design better performing liquid crystal photonic and display devices.

The book consists of three Chapters, two of which cover the fundamental aspects of LC physical properties and electrooptical modes useful for photonics applications, while the last Chapter is devoted solely to new LC photonics and emerging display applications.

The first Chapter describes basic LC properties, important for applications, such as dielectric, optical and visco-elastic behavior, as well as LC cell preparations and surface alignment. A special attention is paid to the new photoalignment technology. Photoalignment possesses obvious advantages in comparison with the usually "rubbing" treatment of the substrates of LC display cells. New advanced applications of LC photonics devices become available with a high spatial resolution even on curved and flexible substrates as well as photonic holes, where the traditional rubbing LC alignment is not possible.

The second Chapter provides a comprehensive review of LC electrooptical properties useful for LC photonics applications. The Chapter enlists all basic LC electroptical modes such as electrically controlled birefringence (ECB), twist (TN) and supertwist (STN) modes. A special attention is paid to the new fast nematic LC modes, such as stress splay twist (SST) mode and linear flexoelectric effect in nematic and cholesteric LC. Fast switching high resolution ferroelectric liquid crystal (FLC) modes such as: deformed helix ferroelectric (DHF), electrically suppressed helix (ESH) and orientational electro-optic Kerr effect were considered in detail as good candidates for the new generation of photonics and display LCD with a better response time than the usual nematic LCDs. A particular consideration of nematic surface bistable modes, including bistable TN (BTN), zenithal bistable mode (ZBD) and optically rewritable (ORW) nematic mode as well as bi and multi-stable FLC mode was made in details. The modes can be used for a new generations of green LC photonics and display devices with extremely low power consumption. Finally various switchable LC diffraction gratings were described and analyzed for the application in beam steering devices, switches, lenses and other LC photonics elements. The Chapter also includes a general review of passive and active matrix addressing schemes for nematic and ferroelectric LC.

The third Chapter highlights the new applications of LC in photonics. The author tries to enlist only the main photonics LC applications and definitely does not pretend, that the book covers all LC achievements in the field. We describe LC photonics applications in fiber optics including switches, polarization rotators, attenuators, polarization controllers, phase emulators and other fiber optical components. Application of nematic and ferroelectric LC for high speed communication systems, producing elements that are extremely fast, stable, durable, of low loss, operable over a wide temperature range, and that require low operating voltages and extremely low power consumption. A considerable part of the Chapter is devoted to switchable LC lenses and LC sensors, based on photolalignment and photopatterning technology. High resolution LC light converters and q-plates for the orbital angular momentum of the light and local polarization control can be also found in the Chapter. Finally novel field sequential color (FSC) FLC devices with a high resolution, low power consumption and extended color gamut were described as an emerging display application for the screens of portable PCs, mobile phones, PDAs, switchable goggles and lenses as well as FLC microdisplays and pico-projectors.

The book is intended for a wide range of engineers, scientists and managers, who are willing to understand the "hot" topics of LC applications in photonics and displays. University researchers and students, who is specialized in the condensed matter physics and liquid crystal device developments in particular in LC photonics will also find some useful information here.

Prof. Vladimir G. Chigrinov
Department of Electronic and Computer Engineering
Hong Kong University of Science and Technology
Clear Water Bay
Kowloon, Hong Kong
Tel: +852 258-8522,
Fax: +852 258-1485
E-mail: eechigr@ust.hk

Chapter 1

INTRODUCTION

The book was conceived as a continuation of the series of the books devoted to the physics and applications of liquid crystal (LC) devices, and liquid crystal; photonics in particular [1-17]. To the best of our knowledge, this is the first book, which considers LC photonics. The physical and electrooptical properties of LC were reviewed in books written by de Gennes [1], Chandrasekhar [2], de Jeu [3], Bahadur [4] and others [5-17]. The encyclopedia of LCs appears, which collects the articles on basic physical and chemical principles of LCs as well as their application strategies in displays, thermogaphy and some other fields [7]. Electro-optical effects and their applications in LC devices were also discussed in books of Blinov and Chigrinov [6,8], Wu [11,13], Khoo [14]. LC optics and LC displays (LCD) were discussed in the books of Yeh [9], Lueder [10, 17], and Boer [12].

LC devices for Photonics applications is a "hot topic" of research. Such elements begin to appear in photonics market [18-20]. The electrooptical modes used for the purpose included the light polarization rotation, voltage controllable diffraction and fast switching of the LC refractive index [18]. Use of photo-alignment technique makes it possible to develop new LC fiber components. Photoaligning materials used to align LC in superthin photonic holes, curved and 3D surfaces and as cladding layers in microring silicon based resonators [15]. The prototypes of new efficient photonics LC devices, such as optically rewritable LC waveguides and voltage controllable diffraction gratings are envisaged [21]. The polarization controllers, polarization rotators, variable optical attenuators and other passive LC optical elements for fiber communication networks are under way.

At the same time application of ferroelectric LC (FLC) for high speed communication systems, producing elements (such as switches, attenuators, polarization rotators) that are extremely fast, stable, durable, of low loss, operable over a wide temperature range, and that require small operating voltages and extremely low power consumption when using bistable and multi-stable switching with optical memory. Passive FLC elements for fiber optical communication systems (DWDM components) based on liquid crystal (LC) cells can successfully compete with the other elements used for the purpose, such as micro electromechanical (MEM), thermo-optical, opto-mechanical or acousto-optical devices.

The prototypes of the novel photoaligned LC devices may include field sequential color (FSC) LC with a high resolution, low power consumption and extended color gamut, which can be used in the screens of portable PCs, mobile phones, PDAs. The switchable goggles and lenses based on new FLC prototypes can be efficiently applied in the new generations of switchable 2D/3D LCD TV. FSC LC microdisplays, which is now one of the most advanced technologies for pico-projectors can be also made. The modulators and switches based on fast responded FLC are envisaged in DWDM elements of optical communication systems, such as switches, filters, attenuators, equalizers, polarization controllers, phase emulators.

The aim of the book is highlighted below:

- to describe the physical properties of liquid crystals and preparation of liquid crystal cells the most important for applications;
- to show how to control the liquid crystal behavior in electric fields by varying its macroscopic physical parameters, cell geometry and addressing schemes;
- to summarize basic electrooptical phenomena, as a basis for liquid crystals devices;
- to introduce new liquid crystal technologies, such as photoalignment and photopatterning;
- to show new photonics applications of liquid crystals: LC photonics elements for passive optical network systems [22], such as LC switches, polarization rotators, variable optical attenuators, photonic crystal/liquid crystal devices;
- to describe new LC photonics devices such as lenses, sensors, light converters, field sequential color LCD.

The book is intended for a wide range of engineers, scientists and managers, who are willing to understand the "hot" topics of LC applications in photonics and displays. We will highlight liquid crystal physical properties, geometry of liquid crystal cell and characteristics of electrooptical effects to choose and/or to develop liquid crystal devices with optimal parameters.

Special attention will be paid to photoalignment technology for LC photonics and emergent display devices. University researchers and students, who who specialize in the condensed matter physics and liquid crystal device developments will also find some useful information here.

The author is very grateful to A. Srivastava, Y. Lu, S.V. Pasechnik, E. P. Pozhidaev, V.M. Kozenkov, D.A. Yakovlev, A. Murauski, A. Muravsky, O. Yaroshchuk, A. Kiselev, V. Belyaev, Z. Brodzeli, A. Bermak, A. Poon, V. Presniakov, S. Valyukh, J. Ho, X. Li, D. Huang, P. Xu, T. Du, F. Fan, Q. Guo, J. Sun and G. Hegde, who have contributed greatly to our research program. We also owe much gratitude to H.S. Kwok, H. Takatsu, H. Hasebe, M. Schadt for the important information and many useful discussions.

References

[1] De Gennes, PG. *The Physics of Liquid Crystals,* Oxford, Glarendon Press, 1974.

[2] Chandrasekhar, S. *Liquid Crystals,* Cambridge, Cambridge University Press, 1977.

[3] de Jeu, WH. *Physical Properties of Liquid Crystalline Materials,* New York, Gordon and Breach, 1980.

[4] *Liquid Crystals: Applications and Uses,* (edited by Bahadur, B.), Singapore, World Scientific, 1991.

[5] Vertogen, G; de Jeu, WH. *Thermotropic Liquid Crystals, Fundamentals,* Springer Verlag, 1988.

[6] Blinov, LM; Chigrinov, VG; *Electrooptic Effects in Liquid Crystal Materials,* Springer-Verlag, 1994.

[7] *Handbook of Liquid Crystals,* (Edited by D.Demus, J. Goodby, G.W. Gray, H.-W. Spiess and V.Vill), Wiley-VCH, 1998.

[8] Chigrinov, VG. *Liquid Crystal Devices: Physics and Applications,* Artech-House, Boston- London, 1999.

[9] Yeh, P; Gu, C. *Optics of Liquid Crystal Displays*, Wiley, 1999.

[10] Lueder, E. *Liquid Crystal Displays*, Wiley, 2001.

[11] Wu, ST; Yang, DK. *Reflective Liquid Crystal Displays,* Wiley, 2001.

[12] den Boer, W. *Active Matrix Liquid Crystal Displays : Fundamentals and Applications*, Elsevier, 2005.

[13] Yang, DK; Wu, ST. *Fundamentals of Liquid Crystal Devices,* Wiley, 2006.

[14] Khoo, IC. *Liquid Crystals,* Second Edition, Wiley, 2007.

[15] Chigrinov, VG; Kozenkov, VM; Kwok, HS. *Photoalignment of Liquid Crystalline Materials: Physics and Applications,* Wiley, August 2008.

[16] Pasechnik, SV; Chigrinov, VG; Shmeliova, DV. *Liquid Crystals,* Wiley-VCH, 2009.

[17] Lueder, E. *Liquid Crystal Displays, Addressing Schemes and Electro-Optical Effects*, Wiley, 2010.

[18] Vladimir G. Chigrinov, *Liquid crystal applications in photonics, Frontiers of Optoelectronics in China,* Vol.3, pp. 1674-4128 (2010).

[19] http://www.lightwaves2020.com

[20] http://www.vescentphotonics.com

[21] Vladimir G. Chigrinov and Hoi-Sing Kwok, *Photoalignment of liquid crystals: applications to fast response ferroelectric liquid crystals and rewritable photonic devices,* pp. 117-144. In a book: Progress in LC Science and Technology: In Honor of Kobayashi's 80th Birthday, 2012.

[22] http://en.wikipedia.org/wiki/Passive_optical_network

Chapter 2

LIQUID CRYSTALS (LC): BASIC PHYSICAL PROPERTIES

The chapter is devoted to basic physical properies of liquid crystals (LC), such as structure and symmetry of LC, surface alignment and mixture preparations. Dielectric, optical and viscoelastic properties of LCs are reviewed, taking into account their relations with the molecular structure and mixture content. We consider the surface phenomena, including photoalignment and cell preparations, strong and weak anchoring conditions as well as behavior of liquid crystals in magnetic and electric fields.

2.1. STRUCTURE AND SYMMETRY OF LC PHASES

Liquid crystals are fluids in which there occurs a certain order in the arrangement of the molecules. As a result, there is an anisotropy in the mechanical, electrical, magnetic, and optical properties. The basis of the majority of specific liquid crystal electrooptical effects is found in the reorientation of the director (the axis of preferred molecular locations) in the macroscopic volume of the material under the influence of an external applied field. Anisotropy of the electrical properties of the medium (of the dielectric susceptibility and the electrical conductivity) is the origin for reorientation, whereas the dynamics of the process also depends on the visco-elastic properties and the initial orientation of the director of the mesophase relative to the field. The optical properties of the medium, its local optical anisotropy are changed as a result of this reorientation of the director (either occurring locally or throughout the whole of the sample) in the all known electrooptic effects. We shall provide a description of the chemical structure and physical properties of liquid crystals, which are important for applications.

Molecules of a specific shape form liquid crystalline phases. The most typical are rod-like molecules or rod-like molecular aggregates, which give rise to conventional nematic and smectic phases. Conventional nematic liquid crystals formed by rod-like molecules constitute a uniaxial medium, with non-polar symmetry. The constituent molecules rotate (freely or hindered) around both their short and long axes.

Typical structure of the nematic rod-like molecule is shown in Figure 2.1. It consists of the two (less three or four) ring systems (B, *B*) sometimes with lateral substituents (D, *D*), and a linking group C between them. The groups A and *A* are the terminal groups (tails of the molecules).

Common structures of the molecular compounds, used for Twist-Nematic (TN) and Supertwist-Nematic (STN)- Liquid Crystal Displays (LCDs) as well fluorinated ones, typical for Active-Matrix (AM)-LCD applications are shown in Figures 2.2. and 2.3. respectively.

A long-range orientation order and the random disposition of the centers of gravity in individual molecules characterize nematic liquid crystals. The degree of the orientation order is characterized by the order parameter:

$$S=1/2 <3 \cos^2 \theta -1>, \tag{2.1}$$

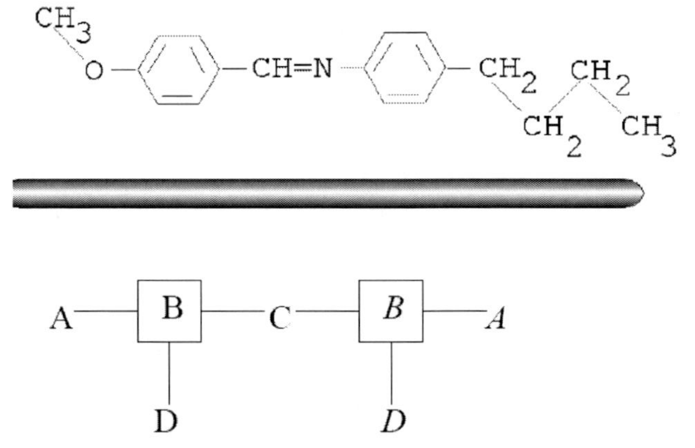

Figure 2.1. Typical rod-like structures of nematic LC molecules.

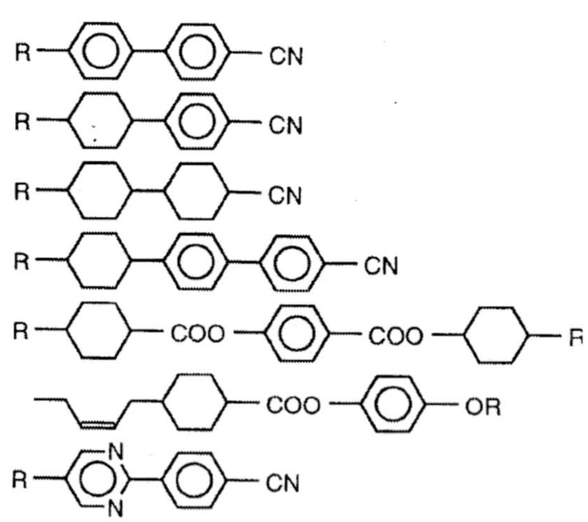

Figure 2.2. The structures of LC compounds for twisted nematic (TN) and supertwisted nematic STN display applications. The typical terminal group $R=C_nH_{2n+1}$, where n=3,5,7.

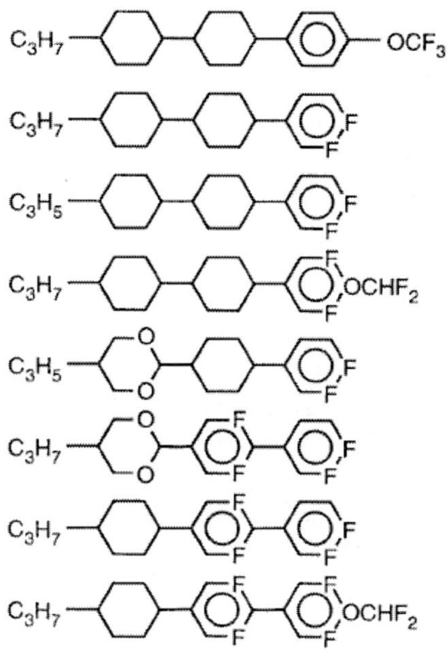

Figure 2.3. Typical structures of fluorinated nematic LC used for active matrix LCD.

where θ is the angle between the axis of an individual molecule and the director of the liquid crystal, and the average is taken over the complete ensemble. In a perfect crystal S=1 or S= -1/2, whereas in the isotropic phase S=0. For nematics S can take, in principle, all possible values $1 \geq S \geq -1/2$. However, for all known nematic phases formed by rod-like molecules the order parameter is positive (negative order parameter would correspond to a nearly perpendicular location of molecular axes with respect to the director).

In principle, one compound may form several liquid crystal phases, the structure of which is defined by a molecular shape (in particular, by molecular symmetry). Being displayed as a function of thermodynamic parameters like temperature, pressure or composition various phases is separated from each other by phase's transition lines. If a series of phase transitions in a liquid crystal occurs over a temperature range, the mesophases are called thermotropic. It is also possible for mesophases to be formed from isotropic solutions of certain materials during the increase of their concentration in a suitable solvent. Such mesophases are termed lyotropic. The nematic (N) phase corresponds to the order parameter S~ 0,7-0,8, which comes to zero in the isotropic (I) phase: S=0. The temperature of the phase transition between N phase and I is called a "clearing point".

In a nematic phase the molecules are statistically oriented along a certain preferred axis **L** called director (Figure 2.4 (a)). The director orientation may change in space, but the characteristic distance of its variation is much longer, than a molecular dimension. By external influence (e.g. through suitable treatment of the walls containing the sample) it is possible to create a uniform orientation of the molecular axes throughout the sample as whole, thus obtaining a liquid monocrystal, or monodomain sample.

Smectic mesophases are characterized by both the orientation and positional order. All of them have a lamellar structure. In smectic phases the liquid crystal molecules are arranged in layers with an average thickness d (Figure 2.4 (b)), which is comparable with a molecular length. The layers can slide freely with respect to each other. In case the molecules more or less perpendicular to the layer normal the liquid crystal phase is called as smectic A. (Figure 2.4 (b)). Optically, the smectic A mesophase, like the nematic one is uniaxial with the optical axis direction coinciding with the director. If the molecules form a certain angle with a layer normal, the liquid crystal phase is recognized as smectic C, chiral or non-chiral dependent whether the rotation with respect to the normal of the molecules in neighbouring layers exists or not. (Figure 2.4(c)). On the temperature scale smectic C phase is usually located lower than smectic A phase. The A-C phase transition is of a continuous type, i.e. of the second order.

When molecules are chiral, i.e. they do not possess the mirror symmetry a variety of chiral mesophases can be observed. One of the examples of these structures called cholesteric or chiral nematic phase is shown in Figure 2.4 (d). Cholesteric liquid crystals are formed by optically active molecules and are characterized by the fact that the direction of the long molecular axes in each successive "layer" (which is made up of molecules which are oriented in parallel and moving freely in two direction) forms a given angle with the direction of the axes of molecules in the preceding layer. In this way a helix is formed, whose pitch (R_0) is dependent on the nature of the molecules (Figure 2.4 (d)). Corresponding to the pitch R_0, the axis of orientation of the molecules (the director) rotates through an angle 2π.

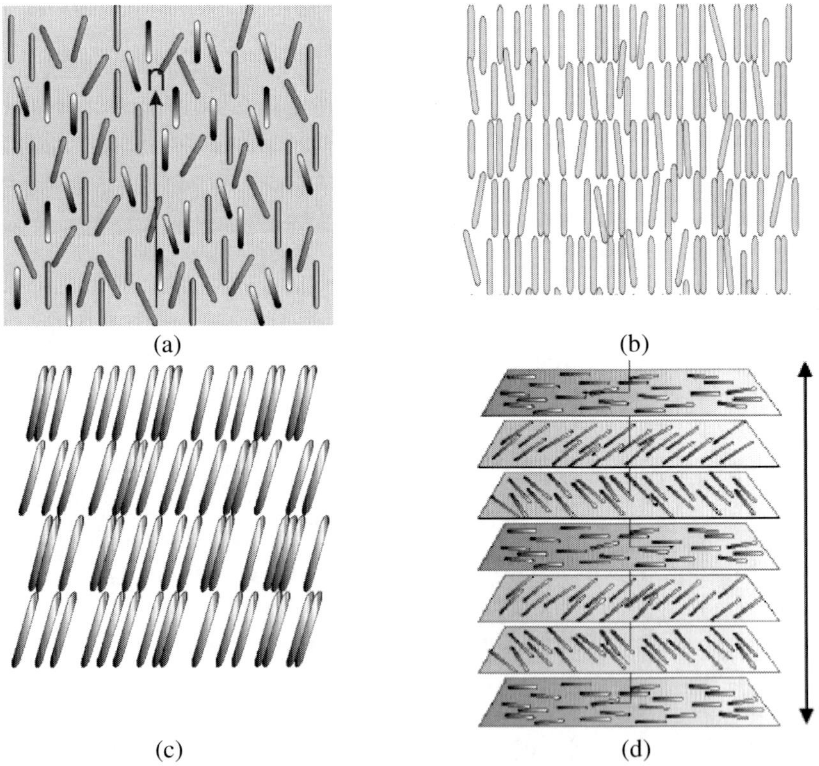

Figure 2.4. Symmetry of LC phases: (a) nematic; (b) smectic A; (c) smectic C; (d) cholesteric.

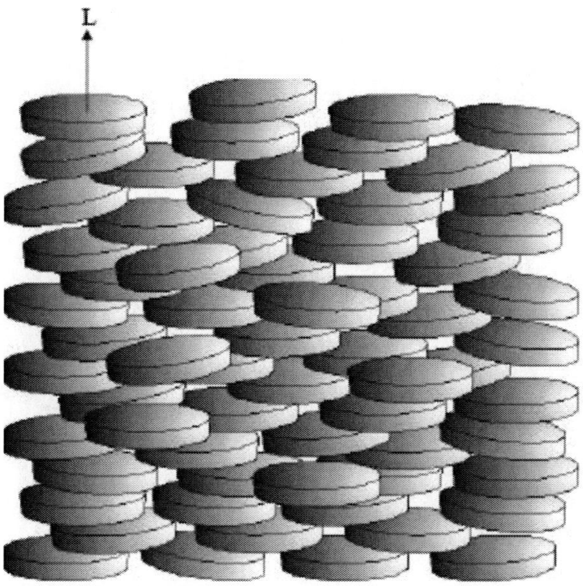

Figure 2.5. Disc-like liquid crystal.

Locally, like nematics, cholesterics are uniaxial. On the macroscopic scale, due to averaging, the helical structure is also uniaxial, the optical axis coinciding with the helical axis, which is always perpendicular to local (nematic) optical axes. Under a microscope cholesterics can show the focal-conic or fingerprint textures. When the helical axis is perpendicular to limiting glasses the uniformly colored planar texture is observed. The color depends on the relative value of the pitch P with respect to a light wavelength.

Disc-like molecules form so-called columnar mesophases (Figure 2.5). This phase is formed in materials with molecules, which are approximately disc-shaped. The discs are packed together in columns, although their arrangement within an individual column can be either ordered or random. The columns themselves can be grouped into hexagonal or orthogonal lattices. As a rule, such a phase is optically uniaxial and negative. The optical axis coincides with the director **L**.

The chiral smectic C phase is formed by optically active molecules, which, in their racemic form, give rise to the conventional smectic C phase. The local symmetry of the C phase (C) is polar since the plane of the molecular tilt (Figure 2.6,a) is no more a mirror plane. Thus, the spontaneous polarization is allowed parallel to the layers (Figure 2.6,b). In the smectic C phase each successive layer is rotated through a certain angle relative to the preceding one so that a twisted structure with the pitch R_0 is formed (Figure 2.6,a). The period of repetition of the physical properties in the C phase coincides with R_0. A classical example of the smectic C phase, which exhibits ferroelectric properties is D (or L)-p-decyloxybenzylidene-p'-amino-2-methylbutyl cinnamate (DOBAMBC) (Figure 2.6,c). Structural classification of the thermotropic liquid crystals most important for applications is provided in Table 2.1 below.

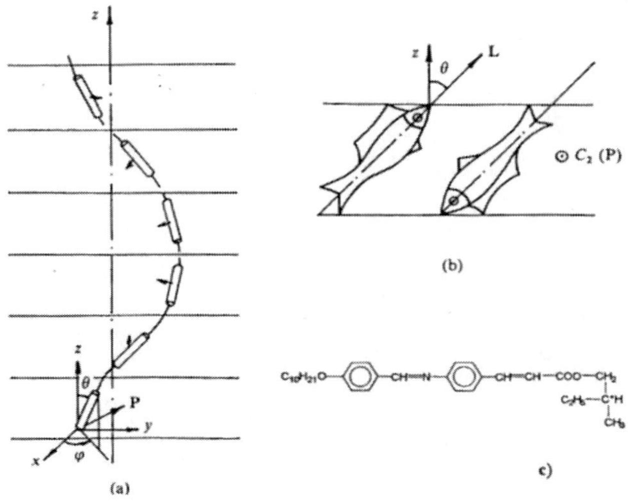

Figure 2.6. Structure and molecular arrangement of a chiral smectic C phase.

Table 2.1. Structural classification of thermotropic liquid crystals

Phase type	Notation	Structure
Nematic	N	The molecules are statistically oriented along a certain preferred axis **L** called director. (Figure 2.4, (a)). The symmetry is not changed by the substitution **L⇒ -L** (Nematic is non-polar in the bulk.)
Chiral Nematic or Cholesteric	Ch or N^*	The axis of orientation of the molecules (the director) rotates through an angle 2π over the distance equals to the helix pitch R_0.
Smectic A	SmA	Molecules are arranged in layers perpendicular to the layer normal.
Smectic C	SmC	Molecules are arranged in layers oblique to the layer normal.
Chiral Smectic C	SmC^*	Molecules are arranged in layers oblique to the layer normal and rotated with respect to the layer normal with a period (pitch R_0). The existence of the spontaneous polarization is possible parallel to the layers.

The important non-liquid crystal compounds are chiral and polar additives, which are used to dope LC material to get highly twisted or polar structures. According to the general assumptions the chiral molecule twists its nearest surroundings, thus inducing the helical rotation in LC cell with a pitch R_0, inversely proportional to the concentration c of the chiral additive. Compounds given below are the examples of the left-handed and right-handed chiral dopants, respectively [1]:

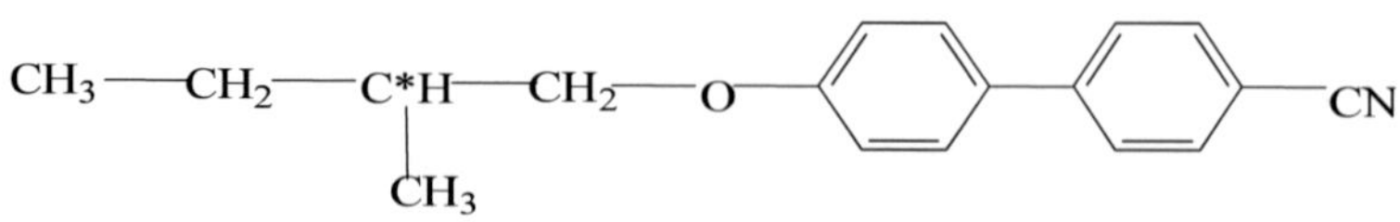

or

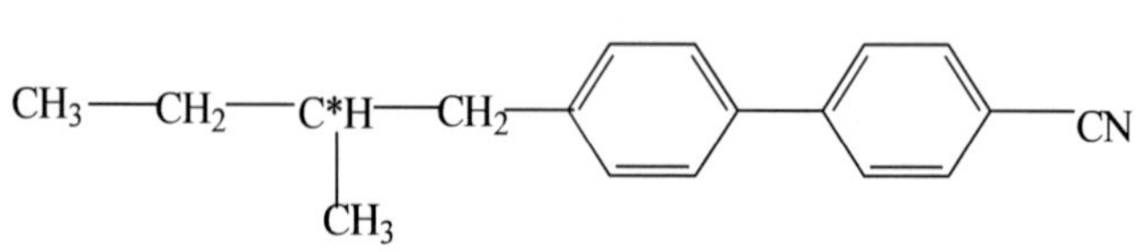

Here the odd-even effect for chirality mentioned in the previous paragraph is used.

Very good results are obtained, using chiral dopants, based on terphenyl derivatives with the structure [2]:

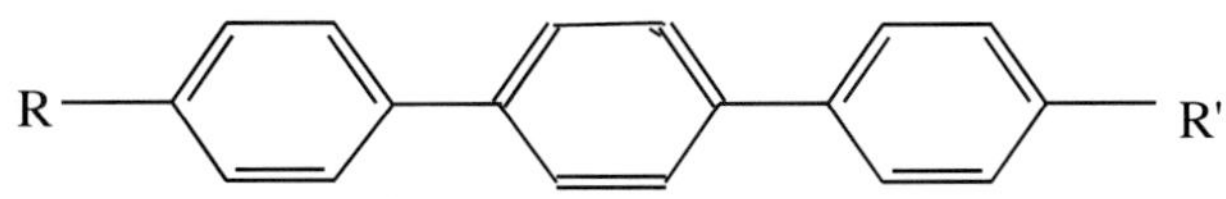

where R is the optically active substituent of the following type:

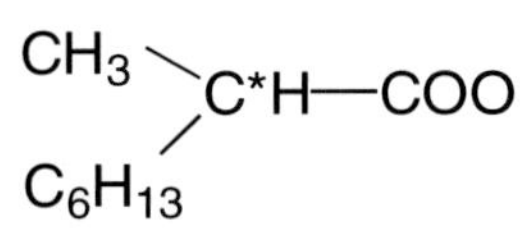

or

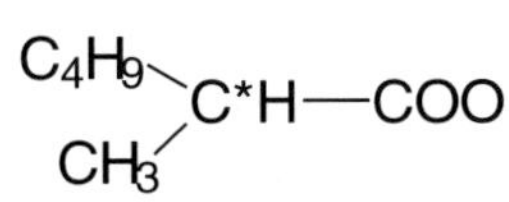

Another very useful compounds are used to form aligning layers for LCDs [1,2]. The aligning layers must satisfy the following requirements: (i) thermal stability; (ii) solubility in suitable solvents; (iii) resistance to the solubility in the LC material; (iv) proper adhesion to the substrate; (v) ability to withstand the high local temperature and mechanical effects of the rubbing or buffing process; (vi) ability to enable the LC molecules pretilt angle on the substrates, which does not degrade at high temperatures. The typical substances used for the aligning layers are polyimide (Figure 2.7 (a)) or polyvinylalcohol (Figure 2.7 (b)) polymers. To make the pretilt angles sometimes the latter are doped by surfactants (Figure 2.7(c)).

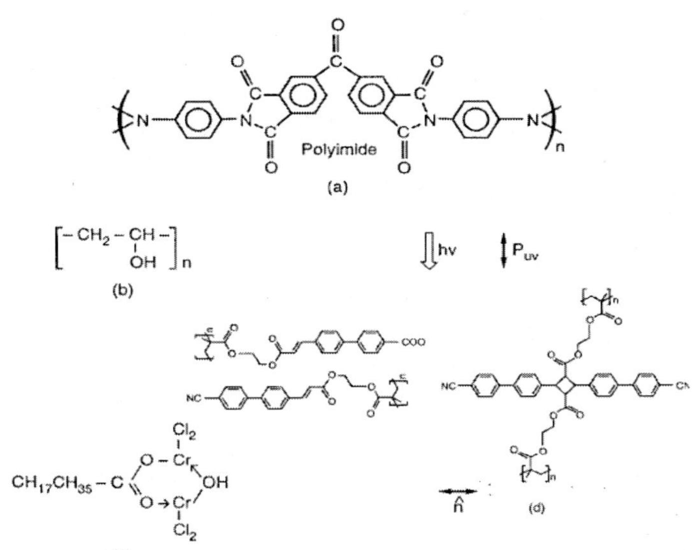

Figure 2.7. Typical liquid crystal aligning materials.

Recently the new “photo-aligning” technique was introduced [3]. Typical materials for this non-rubbing method of the photo-treatment of the substrates are dyes, dyes embedded into polymers or pure photo-polymer films, called also Linear-Photopolymer (LPP) films. One example of LPP films is shown in Figure 2.7(d).

2.2. Typical Methods of Liquid Crystal Materials Preparations for Various Applications

The best liquid crystalline materials for displays are, as a rule, multicomponent mixtures with a wide temperature range of operation. The general principle of constructing liquid crystalline materials for electro-optical applications is to compose multicomponent mixtures satisfying a set of necessary requirements. The requirements are varied for different applications. For example, mixtures for high informative displays must have a specially tailored ratio of elastic moduli, but not extremely high value of the dielectric anisotropy. On the contrary, materials operating at very low voltages (e.g. in image transformers based on semiconductor-liquid crystal structures) require very high values of dielectric anisotropy (for nematics) or high spontaneous polarization (for ferroelectric smectics). However, there are several common requirements for all materials. These are:

- chemical and photochemical stability;
- wide temperature range of operation ;
- low viscosity ;
- optimized electric and optical parameters;
- ability to be oriented by solid substrates.

In fact, the problem of the stability has been solved at least for nematic materials. The majority of compounds mentioned above are chemically stable.

In order to have low viscosity and optimized electric and optical parameters we have to choose carefully compounds from the corresponding chemical classes. For instance, low viscosity and low optical anisotropy are typical of cyclohexane derivatives and the cyano-substituent provides high dielectric anisotropy. Low melting point is achieved by composing eutectic mixtures as it was already discussed. In this case, the odd-even effect ought to be taken into account. To increase the clearing point one has to dope a mixture with a substance having very high transition temperature to the isotropic phase.

This approach can be illustrated with an example of wide temperature range material ZLI-1565 worked out by "Merck". It consists of six components [1, 2] (Table 2.2).

First two components provide the necessary value of the dielectric anisotropy, the second two are introduced to decrease melting point and viscosity and the third pair increases the clearing point. As a result, the mixture has the temperature operating range from -20^0 to $+85^0$ C, viscosity 19 cP (at 20^0C), dielectric anisotropy $\Delta\varepsilon\approx5$-6, and optical anisotropy $\Delta n\approx0.13$.

A special strategy has been developed for composing ferroelectric mixtures. Such materials are based on chiral dipolar compounds which form the smectic C phase. The chemical synthesis of such compounds is extremely difficult. However, the problem can be solved step by step: an achiral smectic C matrix with a wide temperature range may be worked out separately and then doped by a chiral dipolar additive (the chirality and dipole moment can not be decoupled from each other as it was discussed.)

Table 2.2. Wide temperature range material

N	Component	Weight percent age
1.	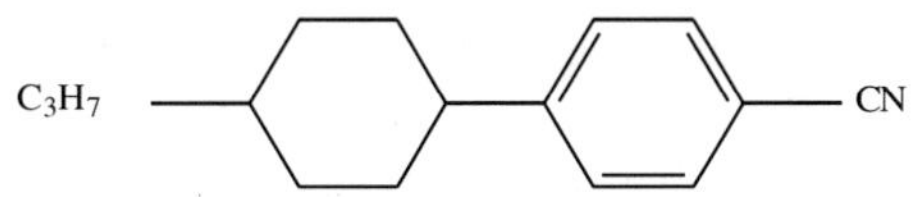	17%
2.	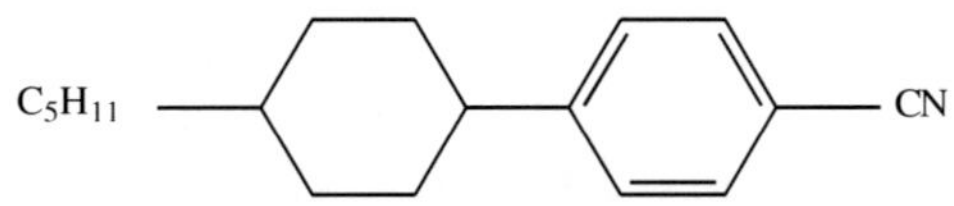	23%
3.	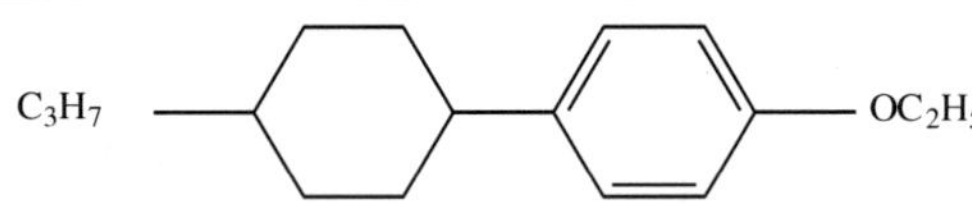	16%
4.	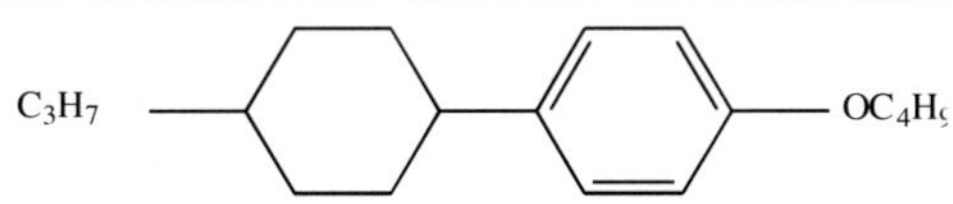	12%
5.	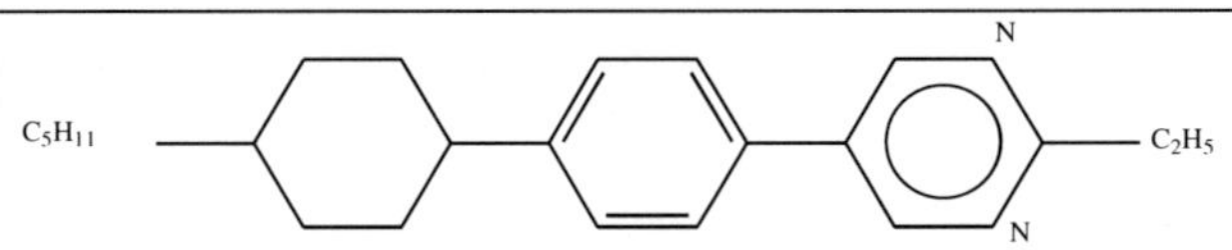	22%
6.	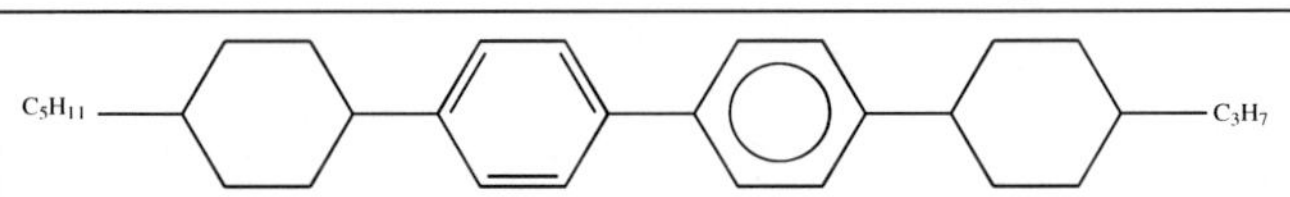	10%

An example of such a mixture which consists of 8 components was described in [1]. A smectic C matrix contains five pyrimidine compounds of general formula:

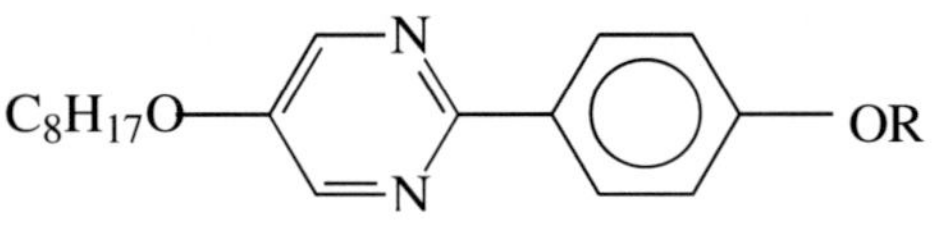

where R= $C_{12}H_{25}$ (8.4 mol.%), $C_{10}H_{21}$ (8.4%), C_8H_7 (4,6%), C_6H_{13} (13.7%), C_4H_9 (15.2%) and 16.7 mol.% of the compound

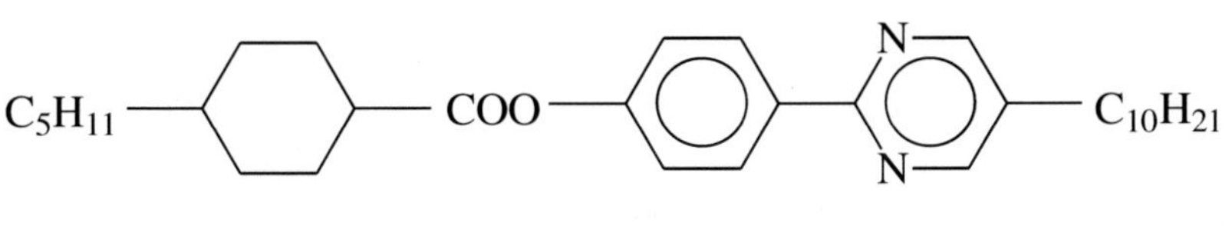

which reduces the viscosity of the material.

The two chiral dopants with a complicated chemical structure not provided here play different roles. The first of them induces strong spontaneous polarization (19, 7%), while the second compensates for undesirable helical structure (1.5%).

As a result the mixture with a wide range (-7.2° - 71.6° C) of the chiral smectic C ferroelectric phase was developed. The value of the spontaneous polarization is fairly high (41 nC/cm^2 at 25° C).

The parameters of the liquid crystal mixtures for active matrix addressing applications are their high purity, which correlates with the value of the electrical conductivity ($\sigma < 10^{-12}$-10^{-14} $Ohm^{-1}cm^{-1}$) and sufficiently high dielectric anisotropy to enable the proper values of the controlling voltages. The best way to fit all these requirements is to use fluorinated compounds, shown in Figure 2.3.

2.3. LC Physical Properties

We will describe the most important physical parameters, which mainly determine electrooptic behavior of liquid crystal cells. According to existent phenomenological theories us first:

(i) introduce these parameters, then;
(ii) illustrate their dependence on the concrete molecular structure; and
(iii) give evidence how to measure them and to develop new liquid crystalline mixtures, having optimal values for them.

We should note that because all the physical properties of the final mixture interconnected and defined by the molecular structure of the components it is impossible to change arbitrary one or another liquid crystal parameter without affecting the others. This is why to develop a new liquid crystal material means a delicate job, the way of certain compromises. The problem seems to be even more complicated because the list of parameters, required for application control is not limited by the physical properties considered in the section. However, a detailed description of quality estimations of the materials is beyond the frameworks of our book. Sometimes this control is even considered as the "Know-How" of the producer.

2.3.1. Dielectric Properties

Pure organic liquids are dielectrics ($\sigma = 0$) and also diamagnetics, so that the magnetic susceptibility, i.e. $\mu = 1+4\pi\chi\approx 1$, while the absorption index $n^2 = \mu\varepsilon\approx 1$.

The value of the dielectric permittivity at optical frequencies ($\varepsilon\ (\omega\Rightarrow\infty) = n^2$ is determined by the average deformation (electronic and atomic) polarizabilities of the molecule ($<\gamma^E>$) through the Lorenz-Lorentz equation:

$$(n^2-1)/(n^2+1) = 4\pi/3\ \rho/m\ N_A < \gamma^E >, \qquad (2.2)$$

where ρ is the density of the substance, m - molecular mass, N_A - Avogadro number,

$<\gamma^E>$ -average polarizability in the electric field E.

The value of ε at low frequencies (the static dielectric permittivity) is determined in the simplest case by the Clausius-Mosotti equation:

$$(\varepsilon - 1)/ (\varepsilon + 2) = 4\pi/3\ \rho/m\ N_A\ (<\gamma^E> + \mu^2 /3kBT) \tag{2.3}$$

where μ^2 /3kBT is the orientation component of the average static polarizability, which depends upon the size of the dipole moment μ of the molecule. Figure 2.8 shows the variation with frequency of the dielectric permittivity for liquids with polar (curve 1) and non polar (curve 2) molecules.

From this must be determined the frequency f_D of its decrease, or the corresponding relaxation time τ_D. The possibility of describing the relaxation of the dielectric permittivity in terms of single time constant is based on Debye hypothesis regarding the exponential relationship governing its return to equilibrium in a constant external field (generally, this is not always satisfied). Based on the general theory of the linear response of the dielectric medium in an external field in which a complex dielectric permittivity describes the phase lag between the displacement D = ε* E and the external field E in the dispersion region of orientation polarizability, Debye's phenomenological equations for the frequency dependence of the dielectric permittivity can be obtained:

$$\varepsilon^* (\omega)-\varepsilon (\infty)= [\varepsilon (0)- \varepsilon (\infty)]/(1-i\omega\tau_D), \tag{2.4}$$

where the real ε' and imaginary ε'' parts of the complex dielectric permittivity ε^* are given by:

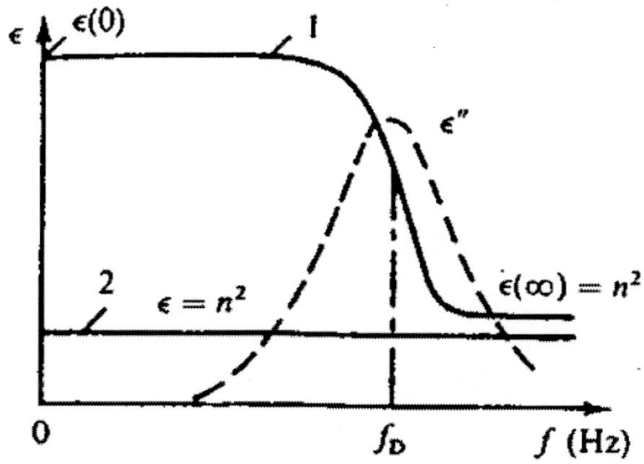

Figure 2.8. Frequency dependence of dielectric permitivities for liquid with polar (curve 1) and non-polar (curve 2) molecules. Dielectric losses are also shown [1-3].

$$\varepsilon' = \varepsilon (\infty) + [\varepsilon (0)- \varepsilon (\infty)]/(1+\omega^2\tau_D^2), \tag{2.5}$$

$$\varepsilon'' = [\varepsilon(0)- \varepsilon(\infty)]\omega\tau_D /(1+\omega^2\tau_D^2)$$

The dielectric losses are determined as follows:

$$\tan\varphi = \varepsilon''/ [\varepsilon' - \varepsilon(\infty)] = \omega\tau_D \tag{2.6}$$

Thus, equation (2.6) describes the curve in Figure 2.8, which represents the frequency dependence of the real component of ε^* (the pure dielectric component), and the characteristic frequency $f_D = \omega/2\pi = (2\pi\tau_D)^{-1}$ .The frequency dependence of the dielectric losses, i.e. the imaginary part of ε^* is also shown in Figure 2.8. These dielectric losses give rise to an active component of the electric current even in a purely insulating medium where there are no free charge carriers. The magnitude of the electrical conductivity which is caused by dielectric losses is provided by the relationship $\sigma_D = \varepsilon''\omega /4\pi$, so that the expression for the complex dielectric permittivity can also be written as:

$$\varepsilon^* = \varepsilon + i\, 4\pi\, \sigma_D /\omega \tag{2.7}$$

A frequency dependence of the average dielectric constant measured parallel to the long molecular axis $\varepsilon_{\|}$ is characteristic for nematic liquid crystals as it corresponds to the polarization contribution related to the molecules rotation along their short axes. The average dielectric constant perpendicular to the molecular axis $\varepsilon_{\perp}$ on the contrary is almost independent in this frequency range as the characteristic times of the molecular rotations along a long axis are several orders of magnitude shorter.

A particularly interesting case occurs when the static value of $\varepsilon_{\|}$ exceeds that of $\varepsilon_{\perp}$. In this case, as a result of the low-frequency dispersion in $\varepsilon_{\|}$ at a certain frequency f_0 a change in sign of the dielectric anisotropy of the nematic liquid crystal $\Delta\varepsilon = \varepsilon_{\|} - \varepsilon_{\perp}$ can occur. Sometimes this frequency is low, particularly in the case of nematic liquid crystals which are formed of long three-ringed molecules, e.g. phenylbenzoates, where the barriers to rotation of the molecules around the short axes are particularly high [4]:

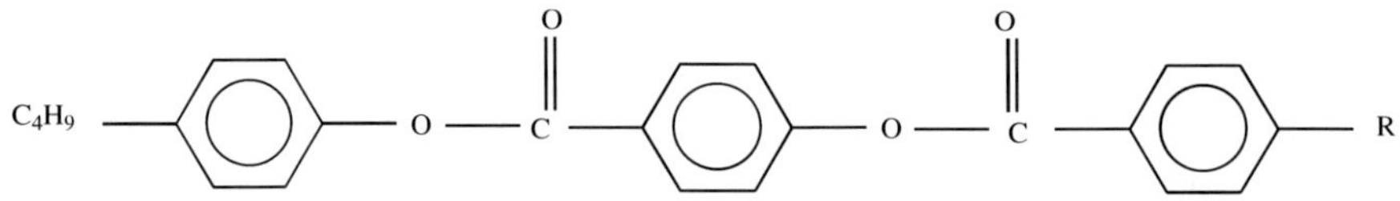

$R_1 = nC_4H_9,\ R_2 = CH_3O.$

Thus, in the binary mixture of these compounds we find $f_0 \approx 20$ kHz at 60^0 C, (Figure 2.9). The dielectric sign inversion frequency f is strongly temperature dependent [1,2,4]:

$$f_0 \approx \exp(-E_0 /k_B\, T), \tag{2.8}$$

where E_0 is the corresponding activation energy. Both the values of f_0 and E_0 are defined by the value of the orientational order parameter and its temperature dependence, thus depend on molecular structure of components in a liquid crystal mixture.

Static values of dielectric constants $\varepsilon_{||}$ and $\varepsilon_{\perp}$ are functions of liquid crystal orientational order S the angle β between the point molecular dipole and the axis of the maximum polarizability of the molecule, the average molecular polarizability $<\gamma^E> = (\gamma_{||}^E + 2\gamma_{\perp}^E)/3$ and its anisotropy $\Delta\gamma^E = \gamma_{||}^E - \gamma_{\perp}^E$. In the frameworks of this theory the liquid crystal dielectric anisotropy takes the form[2,4]:

$$\Delta\varepsilon = \varepsilon_{||} - \varepsilon_{\perp} = 4\pi\rho/M\ N_A\ hF[\Delta\gamma^E - F\ \mu^2/2k_BT\ (1-3\cos^2\beta)]S \quad (2.9)$$

where h and F are taken from the Onsager relationships valid for isotropic dielectrics as functions of the average dielectric constant $<\varepsilon> = (\varepsilon_{||} + 2\varepsilon_{\perp})/3$ and cavity volume $a^3 = 3\mu/4\pi \cdot N_A \cdot \rho$ only. Thus according to (2.9) the temperature dependence of the dielectric anisotropy $\Delta\varepsilon$ is defined only by the temperature dependence of the order parameter S(T) (Figure 2.10).

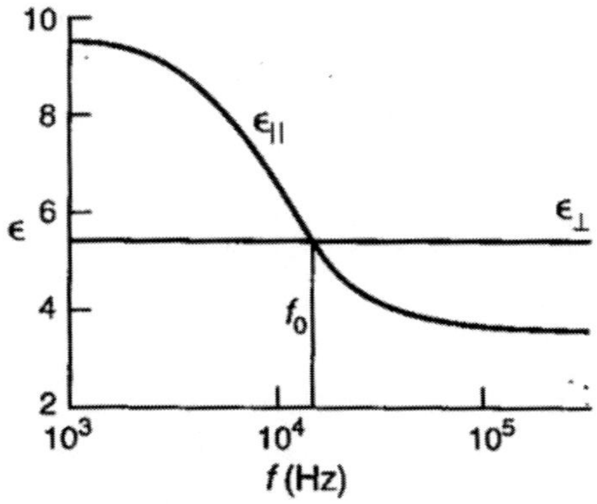

Figure 2.9. Frequency relaxation of the dielectric constant $\varepsilon_{||}$, f_0 is the dielectric inversion frequency [4].

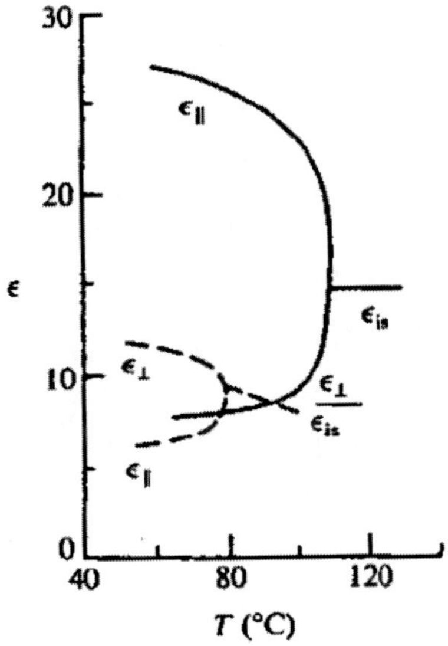

Figure 2.10. Temperature dependence of liquid crystal dielectric constants with positive (above) and negative (below) values of dielectric anisotropy $\Delta\varepsilon = \varepsilon_{||} - \varepsilon_{\perp}$, ε_{is} is isotropic value of the dielectric constant.

In case of isotropic liquid with polar molecules (S=0, $\mu \neq 0$) the well known Onsager equation takes place:

$$\varepsilon_{is} = 4\pi\rho/M\ N_A\ h\ F[<\gamma^E>+F\ \mu^2/3k_BT\ ] \tag{2.10}$$

In view of this for nematic liquid crystals with molecules, which have a large longitudinal dipole moment we have $\Delta\varepsilon > 0$, in good agreement with experiment. Figure 2.10 shows the temperature dependence of the $\varepsilon_{\|}$ and $\varepsilon_{\perp}$ components of two nematic liquid crystals [2]. In the first example, 4-butoxybenzilidene-4'-cyanoaniline

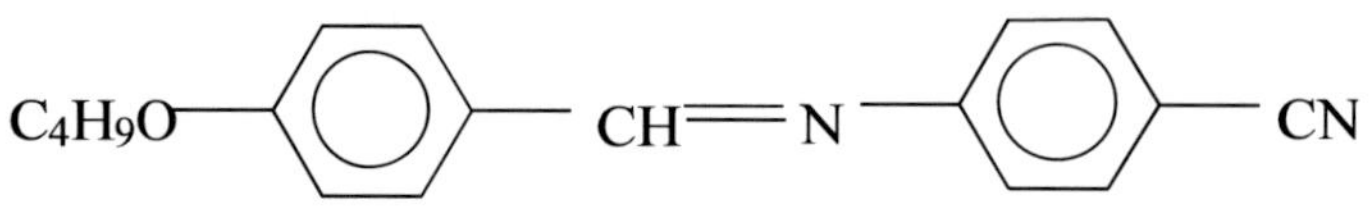

the molecules have a large (ca.4-5 D) longitudinal dipole moment. In the second example, 4-ethoxy-4'-hexyloxy- -cyanostilbene

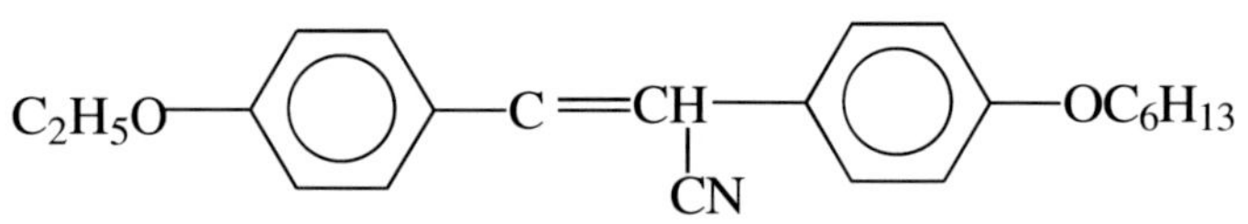

the dipole moment is approximately the same size (because of the nitrile group), but it is directed almost normal to the long axis of the molecule. In the first case the dielectric anisotropy is positive ($\Delta\varepsilon \approx 18$, $T = 80^0$ C), but in the second case it is negative ($\Delta\varepsilon \approx -5$, $T = 60^0$ C).

In liquid crystalline mixtures the following "additivity law" for the effective value of dielectric anisotropy is valid [1]:

$$\Delta\varepsilon_{mixt} = \Sigma_i\ C_i\ \Delta\varepsilon_i \tag{2.11}$$

where C_i is molar fraction of the i-th mixture component, taken at the reduced temperature $\tau = (T_{NI} - T)/\ T_{NI}$ (T_{NI} is the nematic to isotropic transition temperature). According to experiment, the additivity law holds both for weak polar and strong polar mixtures of liquid crystal compounds.

To obtain the desirable value of the dielectric anisotropy in a liquid crystalline material so-called high dipole additives are often used which possess large dipole molecular moment parallel or perpendicular to the long molecular axis. Below two examples of such additives are given [1]:

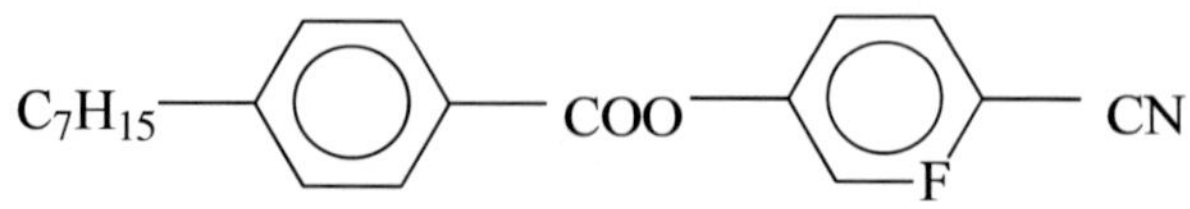

$\Delta\varepsilon = +50$,

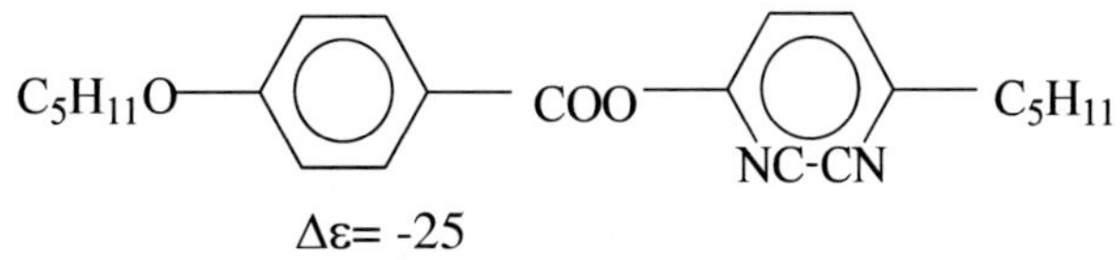

In the first substance the molecular fragments COO, CN and F all contribute to the longitudinal dipole moment. In the second compound two CN groups in the lateral position create a strong dipole moment perpendicular to the long molecular axis. Using high dipole additives it is possible to develop liquid crystal mixtures with wide variation of $\Delta\varepsilon$ values (from -5 to +25 or even wider), required for practical applications.

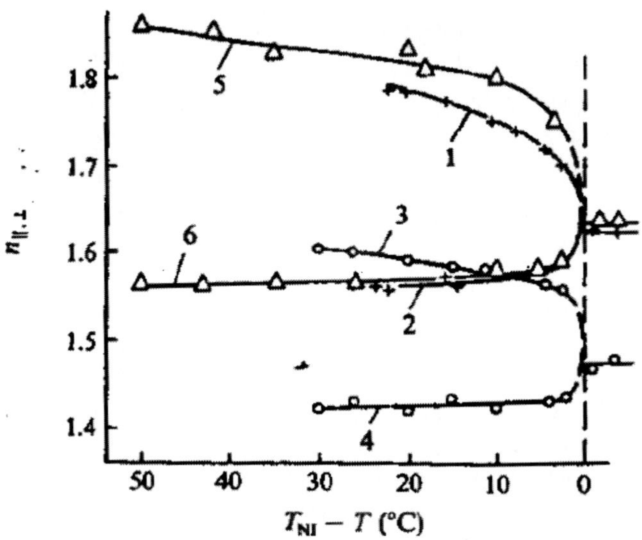

Figure 2.11. Typical dependencies of principal refractive indices $n_{||}$ (curves 1,3,5) and $n_{\perp}$ (curves 2,4,6) for the three liquid crystals with different degrees of conjugation in the linkage groups.

2.3.2. Optical Anisotropy

If we consider the behavior of uniaxial liquid crystals at optical frequencies ($\omega >> \omega_D \approx 10^{10}\ s^{-1}$), the orientation polarization component will not enter into the discussion. A contribution to the electric polarizability of liquid crystal molecules $\gamma_{||}^E$ and $\gamma_{\perp}^E$ at optical frequencies is made only by the electronic and atomic parts. Consequently, both parts of the complex refractive index (that is the refractive index n and the absorption coefficient κ) become anisotropic and each has two principal components ($n_{||}, n_{\perp}$ and $\kappa_{||}$, $\kappa_{\perp}$).

Typical temperature dependences of the principal refractive indices for three nematic liquid crystals are given in Figure 2.11 [2]. According to (2.9) the optical anisotropy $\Delta n(T)= n_{||}(T) - n_{\perp}(T)$ temperature variation is proportional to that of the order parameter S(T).

The average value of the refractive indices in the nematic phase is given by the relationship:

$$<n^2> = 1/3\,(n_{||}^2 + 2n_{\perp}^2) \tag{2.12}$$

with the value $<n^2>^{1/2}$ differing from the refractive index n_{is} in the isotropic phase because of the temperature dependence of the material density. This is due to the fact, that average electrical polarizability of the molecules (as well as its components) are independent of temperature. In Figure 2.11 attention is drawn to the marked difference in the value of the refractive indices for the three materials. This is related to the differences in their molecular polarizability. In accordance with quantum mechanical theory of dispersion, the polarizability of molecule in the ground state (index 0) at frequency ω is proportional to the following sum over all possible quantum transitions from this state to higher states (k):

$$\alpha \approx \Sigma_k f_{0k}/(\omega_{0k}^{2}-\omega^2), \tag{2.13}$$

where f_{0k} and ω_{0k} are the oscillator strength and the frequency respectively of the $(0 \Rightarrow k)$ transitions. If the oscillator strengths of the transitions to the higher levels are approximately the same, then the longest wavelength transitions will make the largest contribution to the polarizability of the molecule at a given frequency, since in this case the denominator of Eqn.2.13 takes its smallest value. In the series of liquid crystal molecules with different linking or bridging groups (Figure 2.1) the longwave absorption band is noticeably displaced to the short-wave region of the spectrum because of the degree of conjugation in the linkage:

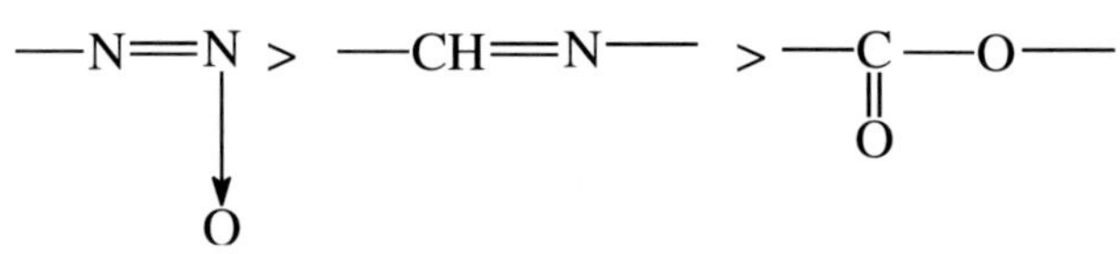

Thus the average polarizabilities of the corresponding liquid crystal molecules decrease (Figure 2.11) and moreover, the frequency dispersion of the refractive index decreases in the same order in agreement with equation 2.13 (Figure 2.12).

In cyclohexanecarboxylic acids the longest wavelength absorption bands are in the far ultraviolet region of the spectrum. Therefore, the optical anisotropy of these compounds is small ($\Delta n \approx 0.05$) and only a small dispersion of the refractive indices is observed.

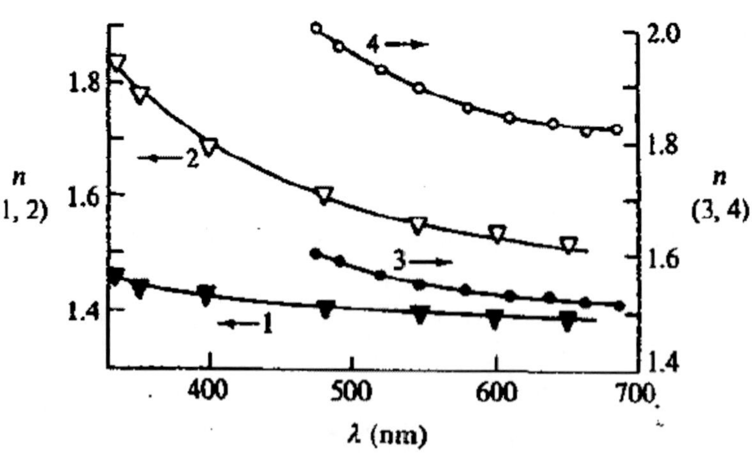

Figure 2.12. Dispersion of the refractive indices $n_{\parallel}$ (2,4) and $n_{\perp}$ for various wavelengths. The two liquid crystals have the conjugation links COO (1,2) and N_2O (3,4) with the structures shown above [1,2].

According to (2.9) the optical anisotropy anisotropy $\Delta n = n_{\parallel} - n_{\perp}$ is completely defined by the anisotropy of polarizability, measured parallel and perpendicular to the long molecular axis.

The values of optical anisotropy Δn increase with [1,2]:

(i) the elongation of the conjugation chain parallel to the long molecular axis;
(ii) replacing saturated aromatic rings to the unsaturated ones;
(iii) shortening the alkyl chain of the end molecular groups in homologues series in form of even-odd alternation;
(iv) increasing values of the order parameter S (2.1). Dependence of Δn on the molecular structure could be illustrated by Table 2.3 below.

Optical anisotropy of liquid crystal mixtures obeys the additivity rule for refractions:

Table 2.3. Optical anisotropy correlation with molecular structure of nematic liquid crystals [1-3]

Physical mechanism	Molecular Structure	Optical anisotropy
(i)	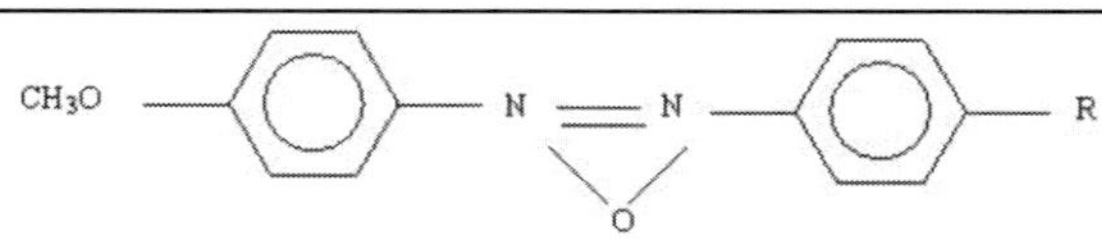 $R=OCH_3$ $R=C_4H_9$	0.26 0.21
(ii)	C_7H_{15} — AR — CN AR= 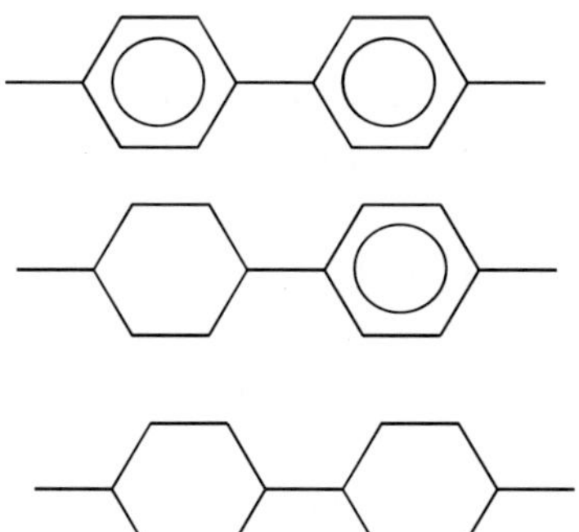	0.16 0.09 0.06

[1] The measurements were made at the wavelength λ=589nm.

$$[1/\rho\,(n^2-1)/(n^2+1)]_{mix} = \Sigma_i C_i\,[1/\rho\,(n^2-1)/(n^2+1)]_i, \tag{2.14}$$

where C_i is a molar fraction of i-th component in the mixture. The validity of (2.14) is confirmed in experiment even for different chemical classes both for $n=n_{\parallel}$ and $n=n_{\perp}$.

2.3.3. Visco-Elastic Properties

The visco-elastic properties of liquid crystals are very important and mainly determine the behaviour of liquid crystals in external electric fields, defining such characteristics as controlling voltages, steepness of the transmission - voltage curve, response times etc.
We shall:

(i) briefly outline the main definitions of the visco-elastic characterization of liquid crystals;
(ii) show the dependence of visco-elastic constants of nematic liquid crystals on the structure and temperature; and
(iii) discuss certain ideas how to develop new liquid crystal mixtures with given viscoelastic parameters.

2.3.3.1. Elasticity

The basic difference between deformations in a liquid crystal and in a solid is that in liquid crystals there is no translational displacement of the molecules on distortion of a sample. This is due to "slippage" between liquid layers. A purely shear deformation of a liquid crystal conserves elastic energy. The elasticity of an isotropic liquid is related to changes in density. In liquid crystals, variations in density can also be characterized by a suitable modulus, but the elasticity which is related to the local variation in the orientation of the director is their principal characteristic.

In the description of elasticity of a nematic liquid crystal the following assumptions are made.

1. Director **L** reorients smoothly as compared with molecular dimension of a liquid crystal. Thus we may conclude, that the order parameter S remains constant throughout the whole volume of a liquid crystal, at a fixed temperature T while only director field **L** varies in accordance with external electric (or some other) fields.
2. The only curvature strains of the director field, which must be considered correspond to the splay, bend and twist distortions (Figure 2.13). Other types of deformation either do not change the elastic energy (e.g., mentioned above pure shears) or are forbidden due to the symmetry. In nematic liquid crystals the cylindrical symmetry of the structure as well as the absence of polarity (head to tail symmetry) must be taken into account.
3. Following the Hooke's low only squares of the director deformations are included into the expression for the free energy.

In view of these suggestions the density of the free volume elastic energy of nematic liquid crystal could be written as:

$$g = 1/2[K_{11}\,(\mathrm{div}\,\mathbf{L})^2 + K_{22}\,(\mathbf{L}\cdot\mathrm{curl}\,\mathbf{L})^2 + K_{33}\,(\mathbf{L}\times\mathrm{curl}\,\mathbf{L})^2] \tag{2.15}$$

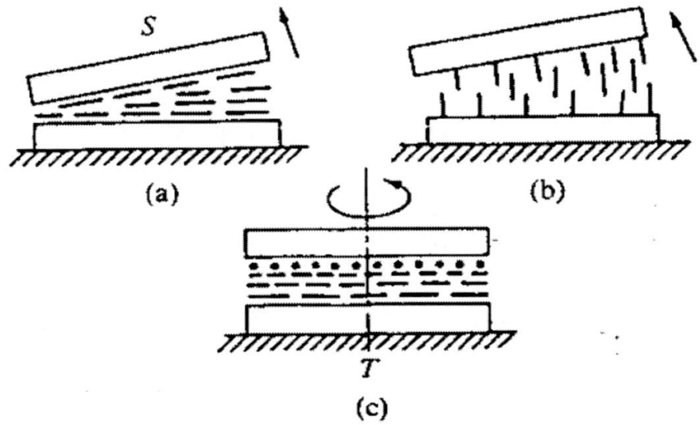

Figure 2.13. Basic types of liquid crystal deformations: (a) S deformation (splay), (b) B deformation (bend), and (c) T deformation (twist).

Equation (2.15) forms the basis for examining almost all electro-optical and magneto-optical phenomena in nematic liquid crystal. The first term in (2.15) describes the S deformation (splay), the second term describes the T deformation (twist) and the third describes the B deformation (bend). These three types of deformation are illustrated in Figure 2.13. Three elastic modules K_{11}, K_{22} and K_{33} in (2.15) characterize the corresponding values of the elastic energies.

In cholesteric (or chiral nematic) liquid crystals the situation is very close to usual nematics. However due to the chirality of the molecules the lowest state of elastic energy in cholesterics does not any longer corresponds to the uniform director orientation, but to the twisted one with a pitch $R_0 = 2\pi / q_0$, where q_0 is the wave vector of a cholesteric. Thus for cholesterics the second term in expression (2.15) must be rewritten as

$$K_{22} (\mathbf{L} \cdot \mathrm{rot}\, \mathbf{L})^2 \Rightarrow K (\mathbf{L} \cdot \mathrm{rot}\, \mathbf{L} + q_0)^2, \tag{2.16}$$

where positive and negative q_0 values correspond to the left and right-handed helix respectively.

The theory of elasticity of smectic liquid crystals has its own features. Deformations related to a change in the spacing between the layers are common to all smectic phases. The deformations are, in general, not related to a change in director orientation, and here an additional modulus of elasticity occurs. In smectic A liquid crystals the only allowed deformation is specific undulation of the smectic layers, such that interlayer distance is kept constant and director remains normal to the layer. This deformation imposes the following limitation to the director field:

$$\mathrm{curl}\, \mathbf{L} = 0 \tag{2.17}$$

and consequently, twist and bend elastic moduli diverge in the vicinity of phase transition nematic - smectic A. The equation (2.17) is not valid for smectic C (and ferroelectric smectic C) liquid crystals, where we also deal with the specific elastic constants.

Since **L** is a dimensionless quantity, the elastic constants K_{ii} must have dimensions of energy/m, i.e. Newton. By dimensional arguments, these moduli should be of the order of

W/a, where W is the energy of interaction of the molecules and a is their size. By assuming W ~ 0.1 eV (a typical value) and a ~ 10A, we derive $K \propto 10^{-11}$ N.

There is a number of experimental data on elastic moduli of liquid crystals, which is only qualitatively explained. The existent molecular approaches do not directly correspond to the real situation, because the molecules are considered as to be spherocylinders or hard rods, what is far from reality [1,4]. For instance, the ratio K_{22}/K_{11} according to the present approaches is about 1/3 which is two times lower that the corresponding experimental range. However, the above mentioned data is quite sufficient for developing liquid crystal mixtures with the required elastic properties.

2.3.3.2. Viscosity

The dynamics of nematic liquid crystals is described by: (i) director field **L** (**r**,t); and (ii) velocities of the centres of the molecules **V**(**r**,t). These variables in general obey the following equations [1-3]:

1. The equation of continuity in incompressible liquids:

$$\mathrm{div}\ \mathbf{V}=0 \tag{2.18}$$

2. Navier-Stokes equation in anisotropic viscous liquid:

$$\rho\,(\partial v_i/dt + \partial v_i/\partial x_k) = f_i + \partial\sigma'_{ki}/\partial x_k \tag{2.19}$$

where $i,k,x_i,x_k = x,y,z$, and

$$f_i = -\,\partial P/\partial x_i + QE_i \tag{2.20}$$

is the external force in the anisotropic liquid dielectric and σ_{ki}' is the viscous stress tensor:

$$\sigma_{ki}' = \alpha_1 n_k n_i A_{mn} n_m n_n + \alpha_2 n_k N_i + \alpha_3 n_i N_k + \alpha_4 A_{ki} + \alpha_5 n_k n_m A_{mi} + \alpha_6 n_i n_m A_{km}, \tag{2.21}$$

P is an external pressure,

$$A_{ij} = ½\,(\partial v_i/\partial x_j + \partial v_j/\partial x_i) \tag{2.22}$$

is analogous to the viscous stress tensor for the anisotropic liquids, and

$$\mathbf{N} = d\mathbf{L}/dt - ½[\mathbf{L} \times \mathrm{curl}\ \mathbf{L}\,] \tag{2.23}$$

is the rate of motion of the director **L**, which vanishes when the entire fluid is in uniform rotation with the angular velocity ½ curl **v**.

The coefficients of the proportionality between viscous stress derivatives and the time derivative of velocity in (2.21) are called viscosity coefficients:

$$\alpha_1, \alpha_2, \alpha_3, \alpha_4, \alpha_5, \text{ and } \alpha_6$$

It can be shown, that

$$\alpha_2 + \alpha_3 = \alpha_6 - \alpha_5,$$

i.e. only five independent viscosity coefficients exist.

3. The equation of the director rotation in a nematic liquid crystal is

$$I d\Omega/dt = [\mathbf{L} \times \mathbf{h}] - \Gamma \tag{2.24}$$

where $\mathbf{\Omega}$ =[$\mathbf{L}$ x d$\mathbf{L}$/dt] is the angular velocity of the director rotation, and I is the moment of inertia for the molecular reorientation, normalized to a unit volume, and

$$\mathbf{h} = -\delta F/\delta \mathbf{L} \tag{2.25}$$

is the functional derivative of the liquid crystal volume free energy with respect to the director components **n** or the so-called molecular field, and

$$\Gamma = [\mathbf{L} \times (\gamma_1 \mathbf{N} + \gamma_2 A\mathbf{L}) \tag{2.26}$$

is the frictional torque, which is analogous to the viscous term in the Navier-Stokes equation.

Here $\gamma_1 = \alpha_3 - \alpha_2$, $\gamma_2 = \alpha_3 + \alpha_2$– viscosity coefficients.

$$\gamma_1 = \alpha_3 - \alpha_2 \tag{2.27}$$

is a so called rotational viscosity of the nematic director **L**, which characterizes the pure rotation of the nematic liquid crystal director without movement of the centers of the molecules, which is called “back flow” effect. The following three viscosity coefficients are important for a proper description of a “back flow” effect:

$$\eta_1 = \frac{1}{2}(-\alpha_2 + \alpha_4 + \alpha_5),\ \eta_2 = \frac{1}{2}(\alpha_2 + 2\alpha_3 + \alpha_4 + \alpha_5),$$

$$\text{and } \eta_3 = \frac{1}{2}\alpha_4 \tag{2.28}$$

Let us, like in previous case, present some known facts about viscosity of a liquid crystal.

1. Temperature dependence of viscosity of isotropic liquids obeys the well known exponential law [1,4]:

$$\eta_{is} = \eta_0 \exp(E/k_B T), \tag{2.29}$$

where E > 0 is activation energy for the diffusion molecular motion, η_0 = const. Similar temperature dependence is observed for all nematic viscosity in the whole mesophase range, except in the vicinity of phase transition regions.

Figure 2.14 shows temperature dependence of ln η_i (2.28) on the inverse temperature 1/T, which are very close to linear functions [1]. As seen from Figure 2.14 the "isotropic" viscosity $\eta_3 = \alpha_4/2$ does not undergo considerable change near the phase transition region, while η_1 and η_2 vanish in the isotropic phase.

The temperature dependence of the viscosity α_i (except α_4) can be expressed in terms of the order parameter S(T) [1,4]:

$$\alpha_i = a_i\, S(T) + b_i\, S^2(T), \tag{2.30}$$

As mentioned above, the most important viscosity combination of Leslie coefficients is $\gamma_1 = \alpha_3 - \alpha_2$, which defines the director response times in electro-optical effects. According to [1,4], the temperature dependence of γ_1 (T, S) could be rewritten in most general cases as

$$\gamma_1 = b\, S^x \exp(E/k_B T) \exp(A/(T-T_0)), \tag{2.31}$$

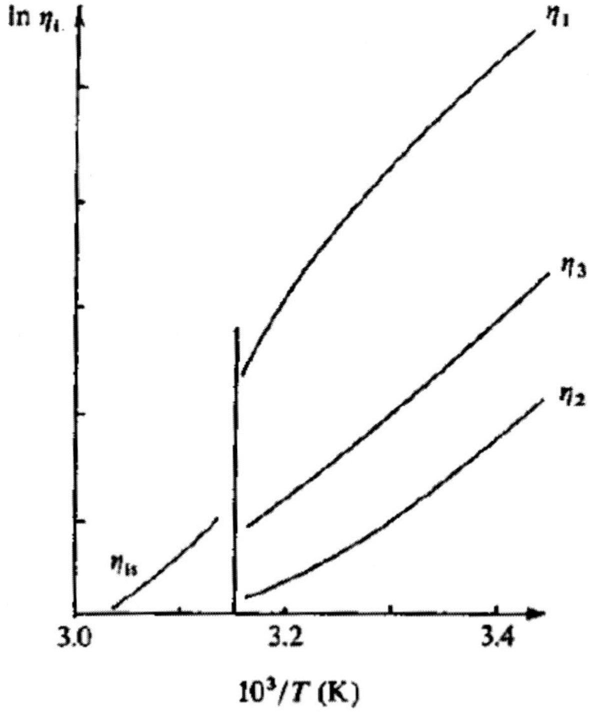

Figure 2.14. Temperature dependence of viscosity coefficients η_1, η_2, η_3,for MBBA liquid crystals [1,2].

where b, A, and T_0 are temperature-independent quantities, which are functions of the molecular structure. For a nematic liquid crystal with latger values of Δn and γ_1, we have x=1, while in the opposite case with low Δn we have x = 2 [2]. The parameter T_0 in (2.31) indicates the so-called temperature of freezing the director motion, which is close to the glass transition temperature T_g when all the dynamic processes are frozen and the viscosity is infinite:

$$T_0 = T_g + Q \tag{2.32}$$

where Q is about 50° K [1].

Finally let us give typical values of rotational viscosity values for different chemical classes (Table 2.4). As seen from Table 2.4, including saturated fragment into the molecular structure, such as cyclohexane ring, results in a visible decrease of γ_1 values and its temperature dependence; however, the Δn value also decreases due to the shortening of the conjugated bonds in the molecular structure. The behavior of viscosity in mixtures of liquid crystal compounds is rather complicated.

The logarithmic additivity law [1,2]

$$(\ln \gamma_1)_{mixt} = x_A(\ln \gamma_1)_A + x_B(\ln \gamma_1)_B \tag{2.33}$$

seems to take place only in the mixtures of homologues in the homologous series (x_A, x_B are molar fractions of the components), while the simple additivity law ($\ln \gamma_1 \Rightarrow \gamma_1$) is not valid at all. A considerable deviation from the relation (2.33) (depression) of γ_1 and E values was observed in the mixtures of strong and weak polar compounds [1]. The physical origin of the phenomenon seems to be similar for elastic constants as discussed above.

There is no theoretical explanation of the viscosity behavior of different liquid crystal substances and their mixtures. Also, there exist only a few works where the viscosity measurements are related to the corresponding molecular structure [1,4]. However, new liquid crystalline low-viscosity materials were successfully developed. To make these materials, the following phenomenological rules should be remembered [1,2].

1. The viscosity is lower for shorter molecules. In homologue series, even-odd alternation is observed with a marked tendency to increasing viscosity with the number of carbon atoms.
2. Alkyl-end groups provide lower values of viscositry as compared with alkoxy- and acyloxy-end groups.
3. Replacement of phenyl ring by a trans-cyclohexane ring results in reduced viscosity values.
4. Introducing the rings with heteroaroms increases viscosity as compared with phenyl analogs (Table 2.4).
5. The most "viscous" bridging groups are the ester one -COO-, the simple bond (as in biphenyls), and the ethane group -CH = CH-.

The most useful compounds for reducing viscosity in liquid crystal materials are cyclohexane derivatives due to their low viscosity, high clearing temperature, and good solubility. We will not consider here the viscous properties of cholesteric, smectic, or some other types of liquid crystals, which is rather complicated, as many unknown viscosity coefficients can be involved in our consideration due to the lower symmetry of these types of liquid crystals.

Table 2.4. Rotational viscosity γ_1 for different LC chemical classes (T=25^{0}C) [1-3]

Structure of the Mixture Molecules*	Rotational Viscosity γ_1, Poise	Activation Energy E, eV	Optical Anisotropy Δn
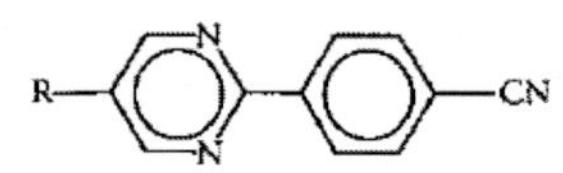	1.9	0.565	0.174
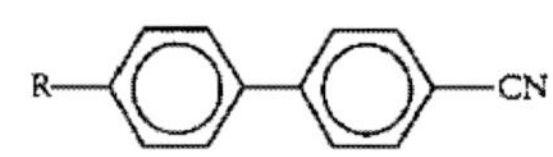	1.1	0.546	0.184
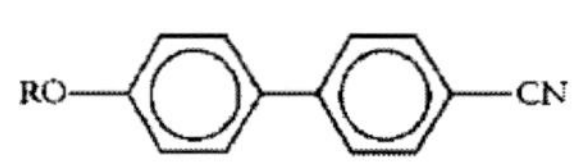	3.3	0.496	0.177
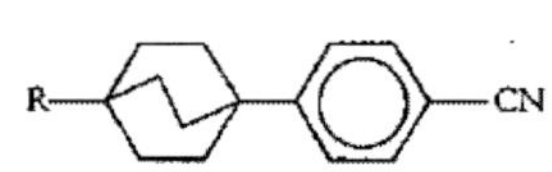	3.6	0.423	0.114
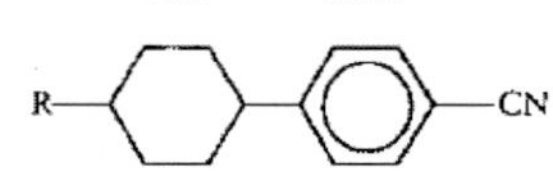	1.0	0.41	0.1

Measurements were carried out in binary mixtures of fifth ($R = C_5H_{11}$) and seventh ($R = C_7H_{15}$) homologues of each compound, taken in the proportion 40:60.

2.4. LIQUID CRYSTAL ALIGNMENT ON THE SURFACE

The interaction of liquid crystals with neighbor phases (gas, liquid, solid) is a very interesting problem relevant to their electro-optical behavior. The structure of liquid crystalline phases close to an interface is different from that in the bulk, and this "surface structure" changes boundary conditions and influences the behavior of a liquid crystal in bulky samples. The nematic phase is of great importance from the point of view of applications in electro-optical devices; thus, in this part we shall concentrate mainly on the surface properties of nematic liquid crystals.

2.4.1. Electrooptic Cells

In most practical applications and when examining liquid crystals, sandwich type cells are used (see Figure 2.15).

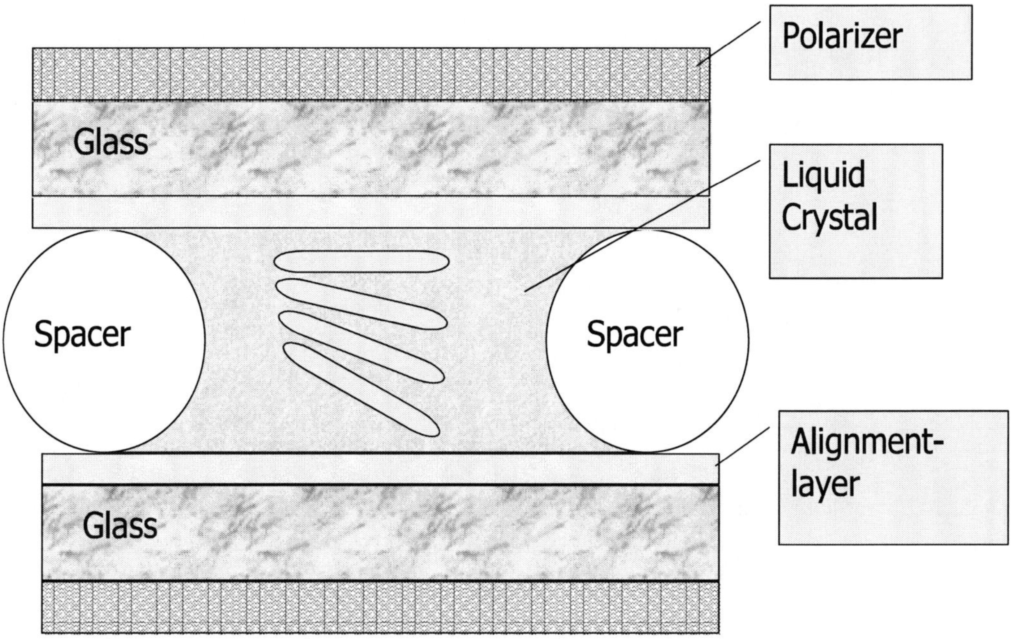

Figure 2.15. Typical liquid crystal sandwich type cell.

A flat capillary with a thickness of 1-10 microns and above is formed from two glass plates with transparent electrodes. The separation between the plates is fixed by means of an insulating spacer (mica, polyethylene, etc.).

To fix a very narrow gap (about 1 micron), glass balls or pieces of glass thread of proper diameter are put between glasses. ITO (indium tin oxide) is coated onto glass usually by sputtering. ITO is conductive, so DC magnetron sputtering can be used with a high efficiency. The sputtering process is performed in a large vacuum chamber and is an in-line process. Typical thickness of the ITO layer is 20-100 nm.

In order to investigate the anisotropy of the properties of liquid crystals and the character of their electro-optical behavior it is necessary to make a definite orientation of their molecules at the boundary walls of the cell.

Molecules in the successive layers “attach” themselves to the molecules on the surface layer and the whole sample will become a monocrysral, either ideal or deformed, depending on the orientation of these surfaces. The orientation of the molecules on the surface is characterized by two parameters: the average angle of the molecules to the plane of the surface θ_0 (preferred direction at the surface) and the anchoring energy W= $\{W_\theta, W_\varphi\}$, where W_θ, W_φ are the corresponding anchoring energies for the polar θ-θ_0 and azimuthal φ-φ_0 deviations of the liquid crystal director from the preferred alignment direction $\{\theta_0,\varphi_0\}$ (Figure 2.16). Using the angle θ_0, we can distinguish various orientations: homeotropic ($\theta_0 = \pi/2$), planar ($\theta_0 = 0$), and tilted ($0 \leq \theta_0 \leq \pi/2$).

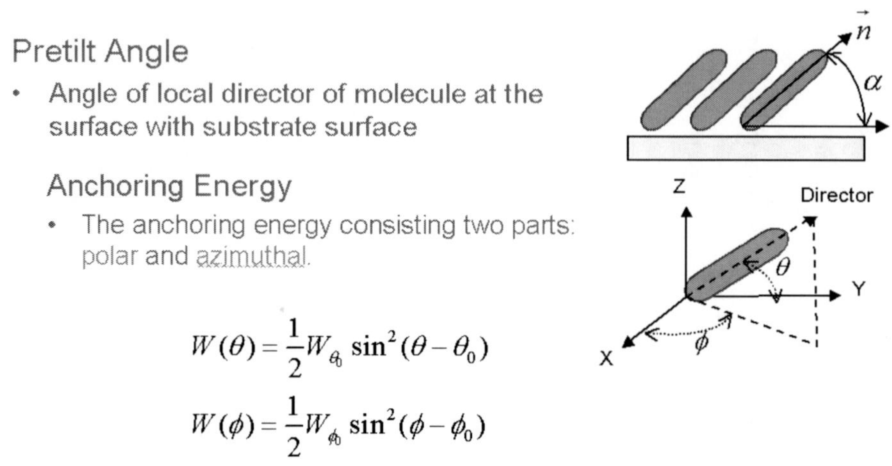

Figure 2.16. Important parameters of liquid crystal alignment: pretilt angle (above) and anchoring energy (below).

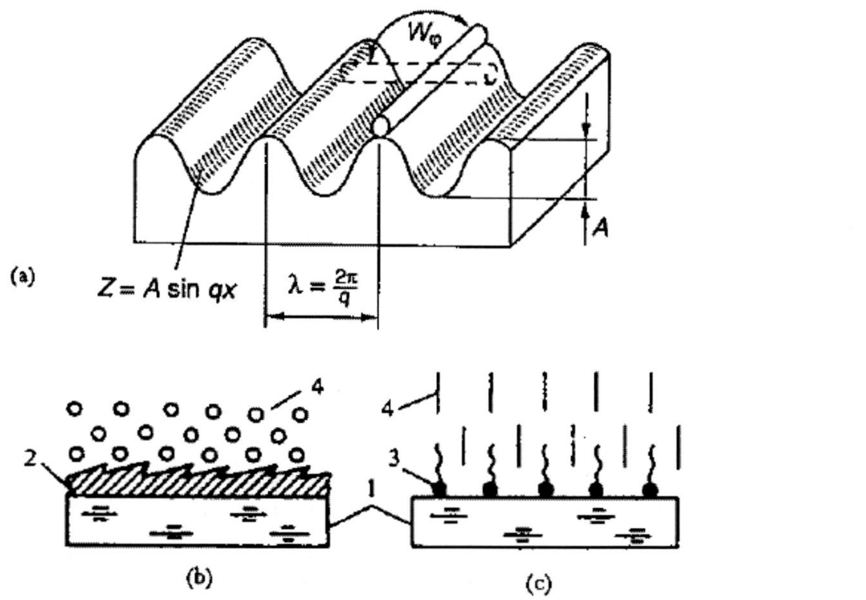

Figure 2.17. Mechanisms of planar (homogeneous) (a, b) and homeotropic (c) orientations of liquid crystals [1]: (a) microrelief obtained by rubbing; (b) microrelief obtained by an obliquely evaporated thin film of metal; (c) orientation of liquid crystal molecules obtained by a surfactant: (1) substrate, (2) obliquely evaporated film, (3) surfactant, (4) nematic liquid crystal. W_φ is the azimuthal anchoring energy [1-3].

2.4.2. Various Types of Liquid Crystal Alignment

2.4.2.1. Planar (Homogeneous) Orientation

Most commonly, a planar orientation is produced by a mechanical rubbing of the surface of the glass with paper or cloth (Chatelain's method). The rubbing creates a microrelief in the electrode coating or glass in the form of ridges and troughs, which promotes the orientation of the molecules along these formations (Figure 2.17). It is very simple, but unfortunately it does not always provide a sufficiently strong anchoring of the director to the surface. Better results are given by the evaporation of metals or oxides (e.g., SiO) onto the surface at oblique incidence. The mechanism of planar orientation of nematic liquid crystals by means of an obliquely evaporated thin film of metal is illustrated in Figure 2.17 (b). If we like to explain why longitudinal ridges and troughs on the surface of the glass promote a planar orientation of a nematic liquid crystals, it is important to study the interaction of a nematic liquid crystal with such surfaces from the viewpoint of a minimum elastic energy [5]. Certain modifications of the Berreman's estimations were proposed recently, but there is still no evident experimental justification of the new approach [6], so we will not consider it in our book. For convenience, the shape of the cross-section perpendicular to the ridges and troughs is taken as sinusoidal:

$$\varphi(z=0) = A \sin qx, \tag{2.34}$$

where x is the direction perpendicular to the rubbing of the surface of the substrate, q is the wave vector of the surface structure, and A is its amplitude (Figure 2.17a). The excess of the elastic energy between the two configurations of the liquid crystal molecules parallel and perpendicular to the grooves (Figure 2.17a) can be calculated by minimizing the nematic liquid crystal elastic energy F in the half-plane (z> 0) with the boundary conditions (2.34) in the form:

$$F = \int_0^\infty g(z)dz = \tfrac{1}{4} KA^2q^3, \tag{2.35}$$

where K is elastic constant of the liquid crystal and $g(z)$ is defined from (2.15). Thus, the additional elastic energy is quadratic, dependent on the depth of the relief (amplitude A) and inversely proportional to the cube of the period of the relief $P= 2\pi/q$. For typical values of the depth and period of the relief ($A=10A^0$, $2\pi/q = 200A^0$) and also of the elasticity modulus ($K=10^{-11}$ N), the value of F is $8{\cdot}10^5$ J/m^2, which is quite reasonable. Thus, this model explains why it is more energetically favorable for the director of a nematic liquid crystal to be aligned along the ridges and troughs of a one-dimensional surface structure (planar orientation). If the relief is two-dimensional (e.g., an etched surface) a similar investigation confirms the advantage of a homeotropic orientation. Monomolecular Langmuir-Blodgett (LB) films have also been proposed as a means of achieving planar orientation [7]. The mechanism of planar orientation in this case has a "chemical" nature (dispersion and polar interaction between the molecules), although a contribution from a purely "elastic" (steric) mechanism is not excluded. Stable monolayers, in which the molecules have their long axes lying on the surface (as in certain organosilicon compounds), are also conducive to planar orientation in a nematic crystal. The quality of the LB film is high in microscopic regions but this is difficult to obtain over macroscopic areas. This is a problem in practical production. The important point is how to decrease surface undulations. Langmuir-Blodgett films are interesting for research in surface material science and the surfaces of the films have many unknown properties [7].

The photo-aligning technique proves to possess visible advantages in comparison with the usually applied "rubbing" treatment of the substrates of the Liquid Crystal Display (LCD) cells. Possible benefits for using this technique include [8]:

1. Elimination of electrostatic charges and impurities as well as mechanical damage of the surface;
2. A controllable pretilt angle and anchoring energy of the liquid crystal cell, as well as its high thermo and UV stability and ionic purity;
3. Possibility to produce the structures with the required liquid crystal director alignment within the selected areas of the cell, thus allowing pixel dividing to enable new special LC device configurations for transflective, multi-domain, 3D and other new display types;
4. Potential increase of manufacturing yield, especially in LCDs with active matrix addressing, where fine tiny pixels of a high resolution LCD screen are driven by thin film transistors on a silicone substrate;
5. New advanced applications of LC in fiber communications, optical data processing, holography and other fields, where the traditional rubbing LC alignment is not possible due to the sophisticated geometry of LC cell and/or high spatial resolution of the processing system;
6. Ability for efficient LC alignment on curved and flexible substrates.

Various mechanisms of photoalignment are described in detail in Ref. [8]. We can distinguish the following mechanisms of the photoalignment (Figure 2.18): (i) photochemical reversible cis-trans isomerization in azo-dye containing polymers, monolayers and pure dye films; (ii) pure reorientation of the azo-dye chromophore molecules or azo-dye molecular solvates due to the diffusion under the action of polarized light; (iii) topochemical crosslinking in cinnamoyl side-chain polymers; (iv) photodegradation (photodestruction), e.g. in polyimide materials. The perfect homogeneous planar and twisted liquid crystal alignment was achieved successfully both in glass and plastic substrates [8].

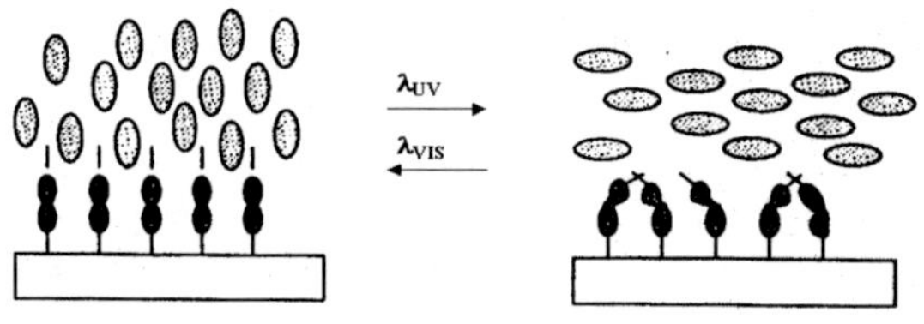

(a)

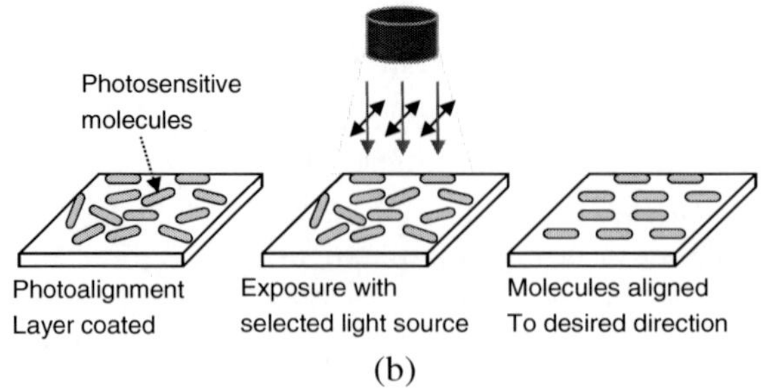

(b)

Figure 2.18. (Continued).

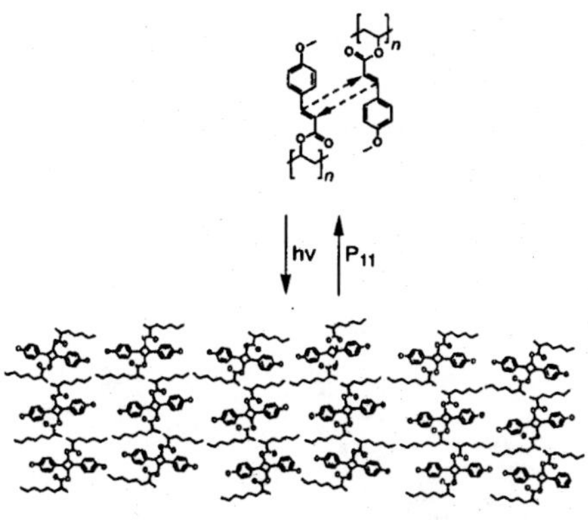

(c)

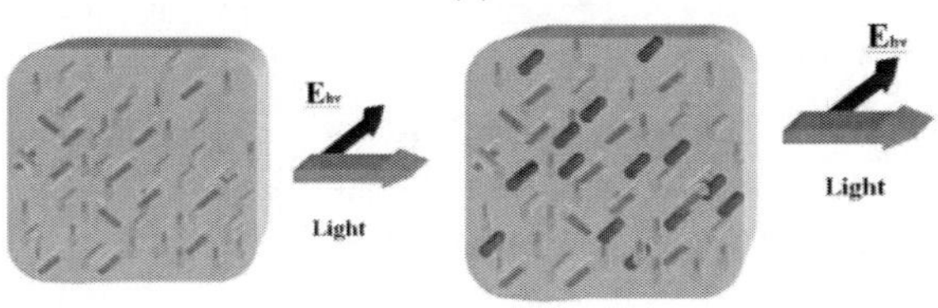

(d)

Figure 2.18. Mechanisms of liquid crystal photoalignment. (a) cis-trans isomerization of the "command layer" attached to the surface; (b) pure diffusion reorientation of the azo-dye layer attached to the surfaceperpendicular to light polarization; (c) photo-crosslinking of the cinnamon fragments, leading to the preferred orientation of the aromatic rings; (d) photodegradation (photodestruction), i.e. selective destruction of the molecular fragment parallel to the light polarization.

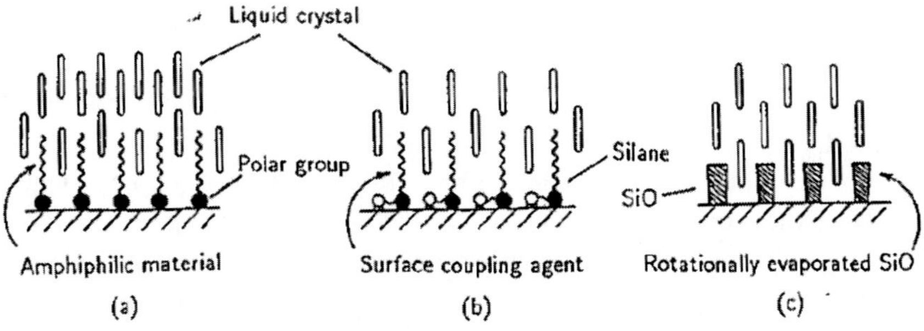

Figure 2.19. Mechanism of perpendicular alignment for a surface with strong polarity (a), for a surface treated with surface coupling agents (b), and for a SiO-rotationally evaporated surface (c) [7].

2.4.2.2. Homeotropic Orientation

There may be three mechanisms for perpendicular alignment as shown in Figures 2.17 (c) and 2.19 [1,7] . The first one is the amphiphilic material assisted alignment, i.e. amphiphilic material adsorbs perpendicularly to the polar surface and the liquid crystal aligns with respect to the amphiphilic material (Figures 2.17 (c) and 2.19 (a)). This amphiphilic material is sometimes an impurity contained in the liquid crystal or surfactant. The second mechanism is the perpendicular alignment obtained by a surface coupling agent, e.g. silanes with long alkyl chains as shown in Figure 2.19 (b). The third mechanism is microscopic columnar structure-assisted alignment as shown in Figure 2.19(c) e.g. obtained by SiO-rotatively oblique evaporation [7].

Homeotropic orientation can be also produced by a photoaligning technique [8]. In this case, the sample with a photosensitive layer is illuminated by a normally incident nonpolarized light. The LC molecules tend to orient perpendicular to the light polarization vector, and consequently the only "allowed" preferred direction remains along the light propagation.

2.4.2.3. Tilted Orientation

Orientation of molecules at a given angle to the surface is achieved using layers of SiO produced by oblique evaporation at a very large angle (80-90°) between the normal to the surface and the direction to the source [7]. In this case, quasi-one-dimensional surface structure is achieved that is oriented at an angle $\alpha=\theta_0$ to the surface (Figure 2.16). This induces a tilt of the nematic liquid crystal molecules in the same direction.

Another method of the pretilt angle generation includes a proper rubbing of polymer (polyimide) films. In this case, the chemical structure of the oriented nematic liquid crystals must be taken into consideration [9]. A very important thing is the presence of alkyl-branches in a rubbed polyimide film [7,9]. The absence of the alkyl-branches results in a low pretilt (~2), polyimide films with a low density of the alkyl-branches leads to a medium pretilt (~5), while a high pretilt angle (~20^0) may be achieved for the high density of the alkyl branches. The physical reason for this is evident. The presence of the alkyl branches increases the tendency for a homeotropic orientation, because they work similar to surfactants on the polyimide surface (see Figure 2.17(c)). Tilted orientation of the nematic liquid crystal molecules can also be achieved by using surfactants [1,2]. The pretilt angle irreversibly increases with temperature due to the increase of the flexibility of the polymer side chains. The effect is more pronounced for the larger dielectric constants of liquid crystals [7].

A high pretilt alignment (~20^0) of the liquid crystal molecules can be achieved by an oblique evaporation technique of SiO (Figure 2.17,b). At present, the tilt angle produced by practical rubbing methods, is about few degrees. Hence, a technique that induces higher tilt angles is required.

The idea of nano-domains and the mixing of vertical and homogeneous alignment polyimide materials together to achieve high pretilt liquid crystal angles was developed [10]. Instead of trying to achieve a homogeneous uniform mixture of two different alignment materials, an inhomogeneous alignment layer consisting of homogeneous and homeotropic domains can be used to achieve the same goal (Figure 2.20). Since it is difficult to fabricate homogeneous mixtures of two different materials, the two materials were simply let to segregate and form domains. As long as the domains are small enough, the result will be a uniform pretilt angle for the liquid crystal molecules. It turned out that reliable high pretilt angles could be obtained [10,11].

Low and high liquid crystal pretilt angles can be also obtained by a photoalignment technique [8], in particular using a non-polarized obliquely incident light (Figure 2.21).

Azo-dye photoalignment materials can be used for the purpose. A subsequent thermo-polymerization of the azo-dye layers in necessary to stabilize liquid crystal alignment [8]. A high reliable pretilt angle can be also made by a nano-structured surface, similar to shown in Figure 2.20, where the rubbing operation is replaced by obliquely incident UV exposure to arrange a photo-degradation process in a homogeneous polyimide. The reliable temperature stable pretilt angle between 0 and 90° were demonstrated [10,11].

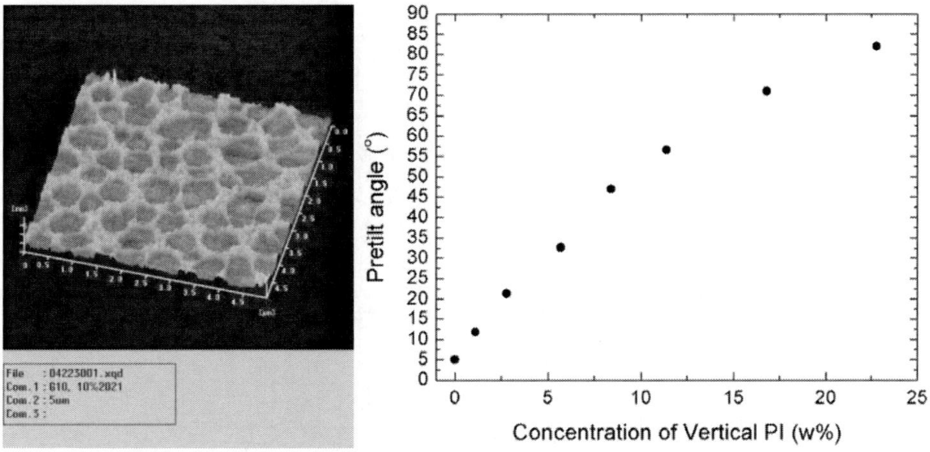

Figure 2.20. Nanostructured liquid crystal surface. Left: Atomic force micrograph of the nanostructured surface. Light (dark) colored region corresponds to vertical (horizontal) polyimide respectively. Right: Measured liquid crystal pretilt angle as a function of concentration of the vertical polyimide [10,11].

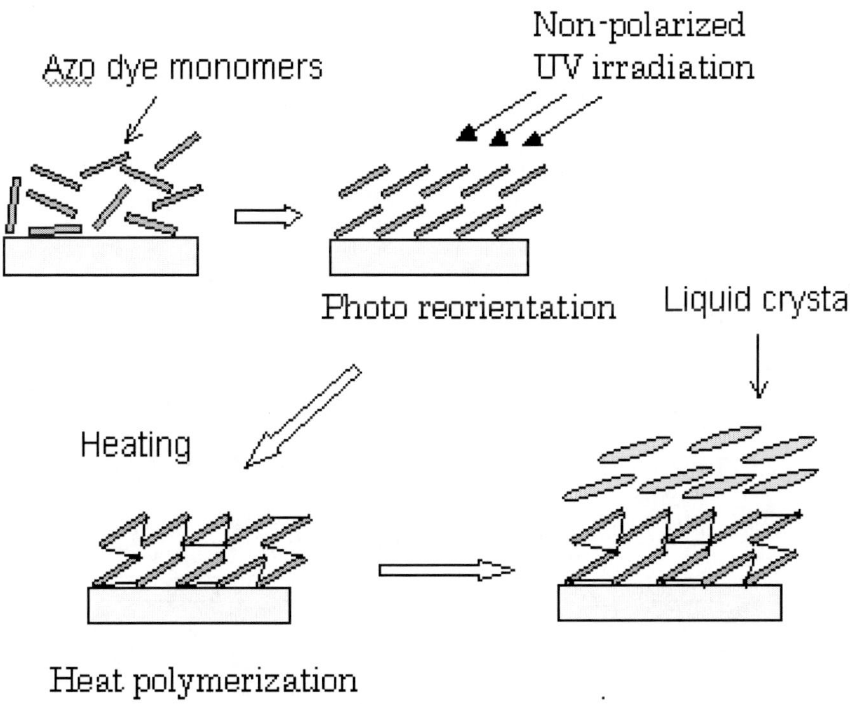

Figure 2.21. Pretilt liquid crystal alignment induced by obliquely incident non-polarized light [8]. The thermo-polymerization stabilized the photoalignment produced by azo-dye layers.

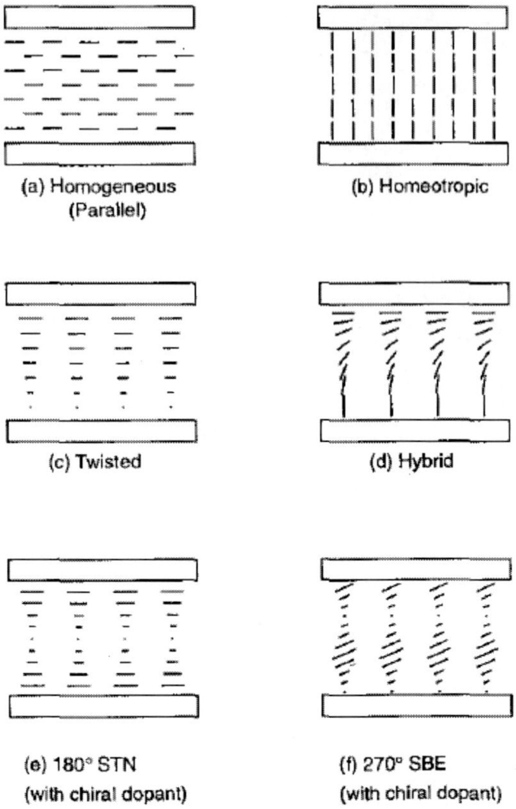

Figure 2.22. Different types of nematic liquid crystal alignment [1].

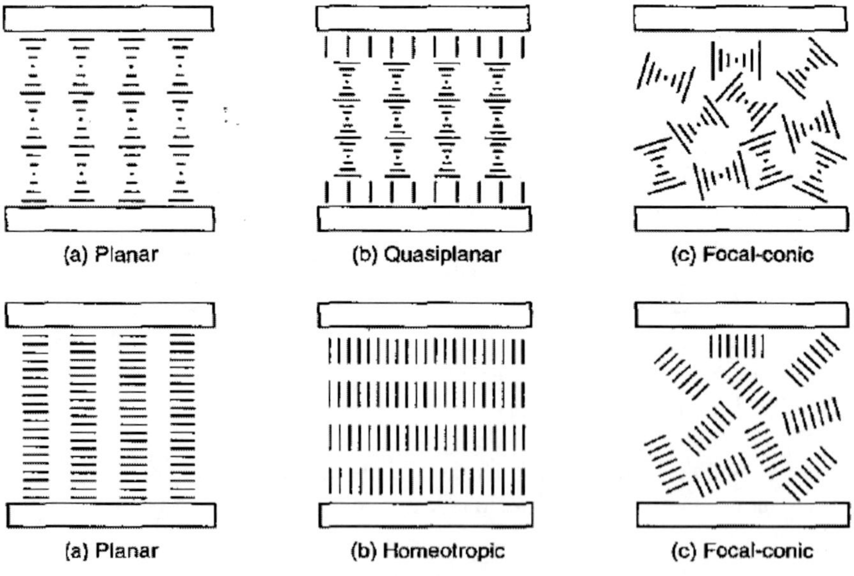

Figure 2.23. Various types of cholesteric (above) and smectic (below) alignment [1].

2.4.2.4. Other Types of Liquid Crystal Alignment

Different types of liquid crystal orientations are shown in Figure 2.22 for nematic liquid crystals and in Figure 2.23 for cholesteric and smectic liquid crystals.

We have already described above the planar (homogeneous), homeotropic, and tilted orientation. Other liquid crystal alignments interesting for applications are of a hybrid (planar to homeotropic transitions between the two substrates) or twisted type (Figure 2.22). We should know that it is impossible to create a twisted nematic liquid crystal orientation with a rotation angle more than 90^0. Thus a certain amount of a chiral dopant has to be added to provide a necessary twist angle of the nematic configuration (180^0 or 270^0 are the most suitable for applications).

The alignment configurations of cholesteric liquid crystals are shown in Figure 2.23. When the director of a cholesteric liquid crystal is oriented along the surfaces of a cell, by their special treatment, the axis of the helix is perpendicular to the glass surfaces. In this case, a transparent, optically active planar texture is obtained. When, on the other hand, the molecules of the cholesreric are oriented normal to the glass surfaces, the axis of the helix is forced to become parallel to the glass, although it can have an azimuthal orientation. In this case, the quasi-planar or "fingerprint" texture takes place. In the more general case, when the directions of the helical axis in the various parts of the cell are random along the azimuth and are not parallel to the glass surfaces, we have a polycrystalline sample-or the so-called focal-conic texture (Figure 2.23). When opposite walls are treated in a different way (with the planar and homeotropic orientation at opposite boundaries), a hybrid cholesteric texture is obtained.

2.4.2.5. Alignment of Ferro and Antiferroelectric Liquid Crystals

An alignment of ferroelectric liquid crystals (FLC) and antiferroelectric liquid crystals (AFLC) can be obtained by special means [1]. The most important factors to consider in order to realize a bistable switching in FLC are (i) to avoid zigzag defects due to the smectic layers tilt in two opposite directions; (ii) to produce an appropriate anchoring energy by rubbing the polymer aligning layer (small energy results in surface defects, while strong anchoring prevents the effect of bistable switching in an electric field [1,2]); (iii) to obtain sufficiently high pretilt angles on the substrates to avoid the degeneration of FLC switching trajectory; and (iv) to prevent the accumulation of the surface charge near the boundaries, which diminish the effective switching field. The latter problem was shown to be solved by orienting an FLC on the photoaligned substrates to diminish the number of the surface traps near the substrates [8]. Perfect electrooptical performance of the photoaligned FLC display with a memorized gray scale was demonstrated [8].

Contrary to nematic liquid crystals there is no standard method for preparing uniform defect-free FLC orientation. In the smectic C* phase both the director and the smectic layers should be oriented in a proper way.

One of the standard techniques often used for this purpose is the preparation of a perfectly aligned smectic A sample. Liquid crystal often undergoes the following phase sequence: isotropic phase ⇒ (nematic) ⇒smectic A ⇒ smectic C. The nematic phase (if it exists) could be oriented by traditional methods, such as the rubbing of polymer films or the evaporation of inorganic films at an oblique angle. The quality of orientation improves for lower values of the specific heat at the nematic ⇒ smectic A transition and the minimum amount of crystallization centers.

If ferroelectric liquid crystal (FLC) material does not possess the nematic phase, the smectic A phase could be oriented directly from the isotropic state. In thick (> 100 μm) cells, and even in thin ones with weak anchoring of the director to boundaries, it is possible to grow well-oriented smectic A layers using either a magnetic field or the temperature gradient. In both cases, the preferred alignment direction is predetermined by a rubbed polymer layer, or by an edge of the spacer fixing the cell thickness and playing the role of a crystallization center [1,2]

FLCs can be aligned using a parallel shift of one cell substrate with respect to the other. The smectic layers tend to orient perpendicular to the substrates ("book-shelf geometry", Figure 2.24), when mechanical shear is applied perpendicular to the rubbing direction [1,2].

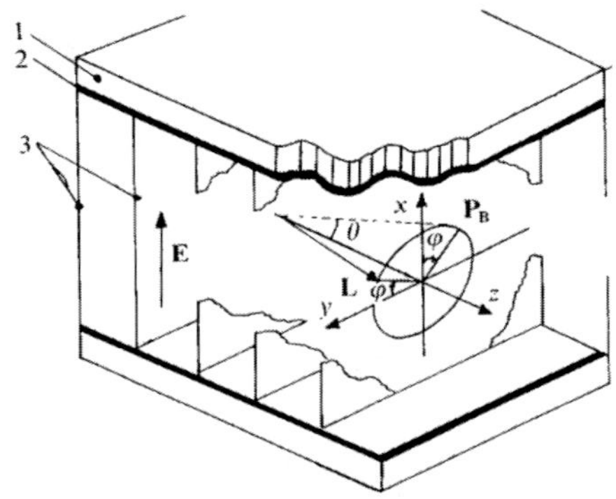

(a)

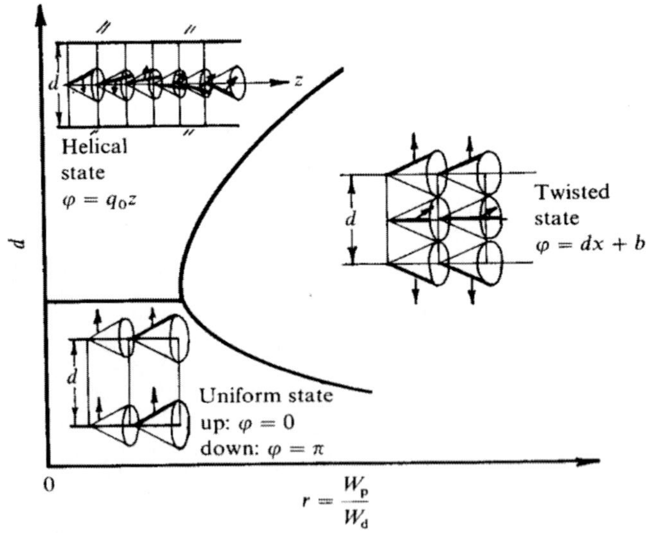

(b)

Figure 2.24. a) FLC texture with smectic layers (3) perpendicular to the substrates (1) with current-conducting ITO layers (2) in "ideal" defectless configuration. (b) Different FLC states in zero field for a various cell thickness d and anchoring ratio $r=W_p/W_d$.

This technique is one of the most reliable and allows us to orient the smectic A phase in a broad range of cell thickness over an area of several square centimeters. FLC mixtures having no smectic A phase can be oriented by cooling down the nematic phase in an a.c. electric field which defines the direction of a smectic layer, while the director is oriented by rubbing the polymer films on the substrates [1,2].

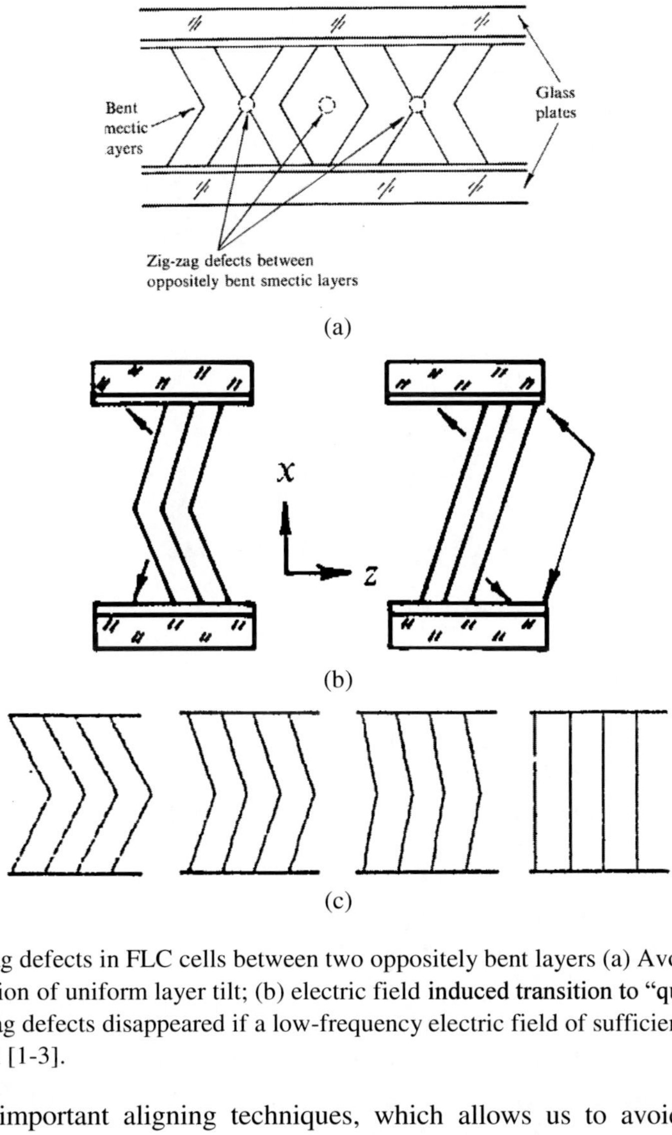

Figure 2.25. Zig-zag defects in FLC cells between two oppositely bent layers (a) Avoiding of zig-zag defects by preparation of uniform layer tilt; (b) electric field induced transition to "quasy- bookshelf" structure; (c) zig-zag defects disappeared if a low-frequency electric field of sufficient amplitude is applied to FLC cell [1-3].

One of the important aligning techniques, which allows us to avoid zig-zag defects between two adjacent smectic layers bent in opposite directions (chevrons), remains the oblique evaporation of silicon monoxide [1,2] (Figure 2.25). Zig-zag defects are avoided by the promotion of only one possible bend or tilt of the smectic layers due to a specific oblique director orientation at the boundaries (Figure 2.25). Samples with the antiparallel evaporation direction contain uniformly tilted layers. Those with parallel directions exhibit the chevron structure of tilted layers free of zig-zag defects.

The chevron structure and zig-zag defects were shown to disappear if a low-frequency electric field of sufficient amplitude is applied to an FLC cell (Figure 2.25, c) [1,2].

The voltage required for the reorientation decreases with increasing spontaneous polarization. The texture appearing resembles the "book-shelf" geometry of smectic layers perpendicular to the substrates. It has minimum response time, maximum transmission difference between both stable states, and a relatively uniform viewing area. Therefore this texture, referred to as "quasi-bookshelf geometry", seems to be very promising for display applications [1,2].

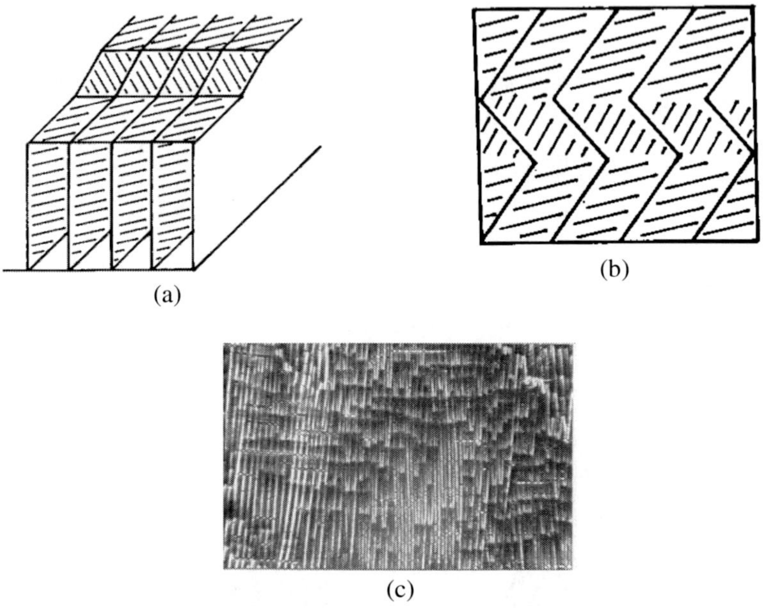

Figure 2.26. Typical "horizontal-chevron" texture for Short Pitch FLCs. (a) SPFLC cell; (b) top view;(c) photograph of the texture in crossed polarizers. (Courtesy by M. Schadt.)[3].

The typical texture for FLCs having a short pitch R_0 in comparison with the layer thickness d (d>> R_0) is shown in Figure 2.26. The structure occurs, when an a.c. electric field (15 V/μm, 10 Hz) is applied during cooling from the isotropic phase down to the smectic A phase and further to the ferroelectric phase [1,2].

This stripe texture means a periodic modulation of the refractive index perpendicular to the lines, while the lines themselves are parallel to the rubbing direction. The spacing between lines increases with increasing the cell gap. The stripes seem to suppress the formation of the FLC helix.

The optical quality of short pitch FLC (SPFLC) cell could be improved if only one substrate is rubbed, and FLC layers are allowed to slide along the other untreated substrate with a low anchoring energy. The final structure will be tilted-bookshelf smectic layer structure (Figure 2.25,b) with a high optical quality.

Photo-induced alignment for FLC cells based on the azo-dye films was investigated [8]. The alignment quality of FLC cells depended on the asymmetric boundary structure and treatment of ITO layer, including photo-aligning substrate and planar substrate. Asymmetric boundary conditions may be useful to avoid a competition in aligning action of solid surfaces of FLC cells. High exposure energy resulted in high-quality of photo-aligned azo-dye layer, and hence, perfect electrooptical response of FLC display. An optimal (about 3–5 nm) azo-dye layer thickness that provided both the highest multiplex operation steadiness and the best contrast ratio of FLC display cells was found [3,8] (Figure 2.27). A prototype of passively addressed FLC display based on the photo-alignment technique has been developed with perfect bistable switching in the multiplex driving regime [3,8].

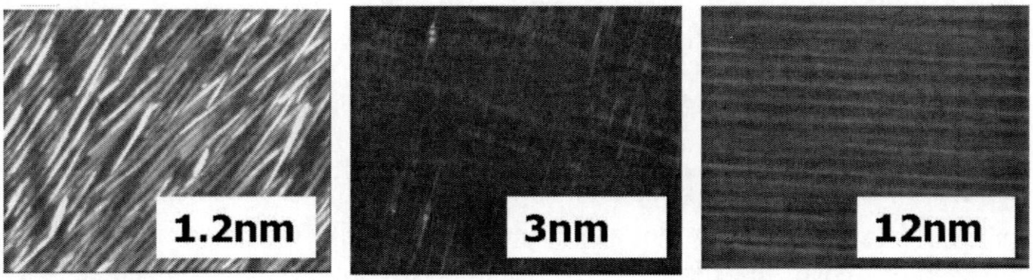

Figure 2.27. Improve of FLC quality by a photosensitive azo-dye layer with an optimal thickness. The optical quality of FLC cell is the best for the azo-dye layer thickness equal to 3 nm [3,8].

In AFLC cells, the optical axis is located neatly along the rubbing direction. Thus the alignment is stable, and is more easy to obtain than in the FLC case. The absence of the macroscopic polarization in the AFLC state, as well as a possibility to avoid surface defects, make the surface AFLC orientation more simple to produce for display applications [1]. The polyimide (PI) photoaligned films were applied for the first time for the alignment of antiferroelectric liquid crystal (AFLC) molecules, which needed planar alignment [12]. Liquid crystal molecules were aligned perpendicular to the polarization direction of UV light. AFLC cell showed some defects, however, the alignment of AFLC cell was enhanced by controlling the cooling rate and post-UV treatment.

2.4.3. Surface Energy

The energy, which is needed to deviate the liquid crystal molecules (the director) from the preferred orientation at the surface, is called anchoring energy. The anchoring energy should not be missed with the energy of adhesion of liquid crystal with the solid surface or the surface energy of the liquid crystal-solid interface, which are of the order of 0.2-0.4 J/m^2, i.e. several orders of magnitude higher than the anchoring energy of the director reorientation at the surface (10^{-6}-10^{-3} J/m^2) [1]. We will define the anchoring energy more in detail.

Assume the interface is in the x, y-plane (see Figure 2.16) and the equilibrium position of the director (the easy direction) is defined by the polar θ_0 and azimurhal φ_0 angles. The easy direction at the nemaric-solid interface is predetermined by a specific treatment of the solid surface. We can distinguish the homeotropic ($\theta_0 = \pi/2$), planar ($\theta_0 = 0$), and tilted ($0 < \theta_0 < \pi/2$) otientation. In turn, the planar orientation can be homogeneous (with a unique angle φ_0, heterogeneous (when several easy directions are possible at the crystalline substrate), or degenerate (all angles are equally probable and the cylindrical symmetry with respect to the surface normal exists). The same is true for the x-y projection of the director for the tilted orientation.

The anchoring energy shows how much energy one has to spend in order to deviate the director from the easy direction (where it is anchored). In the first approximation, the anchoring energy is expressed by the Rapini potential (Fig 2.16) [1,2]:

$$W = -1/2\ W_0\ (\mathbf{L}\cdot\mathbf{d}) \quad (2.36)$$

As the director deviation in the polar and azimuthal plane came to be characterized by different anchoring energies, the terms "polar" and "azimuthal" anchoring energies were introduced. The corresponding contributions to the anchoring were written as (Figure 2.16):

$$W_\theta = ½\ W_{\theta 0} \sin^2 (\theta - \theta_0),\ W_\varphi = ½\ W_{\varphi 0} \sin^2 (\varphi - \varphi_0), \quad (2.37)$$

However, (2.37) does not provide an adequate description of the anchoring phenomena in the case when the liquid crystal alignment angle is high, so instead of (2.37), the more general expression is preferred in the form of the special series-for example, Legendre polynomials [13]:

$$W(\theta) = -1/2 \sum_{n=1}^{\infty} A_{2n}P_{2n}(\cos\theta) = -1/2W_0\cos^2\theta - 1/2 \sum_{n=2}^{\infty} W_{2n}\cos^{2n}\theta \quad (2.38)$$

The shape or the Rapini anchoring potential is shown in Figure 2.28. However, some experimental data may be fit better with other shapes of the surface potential. For example, the elliptic sine shape

$$W = 1/2\ W_0\, \mathrm{sn}^2(\theta - \theta_0, k),\ 0 \le k \le 1 \quad (2.39)$$

has been predicted in [14], which is reduced to the Rapini shape only for a special value ($k = 0$) of the modulus of the elliptic function ($0 \le k \le 1$). This function can describe a very sharp minimum of the surface potential near the easy direction, (see Figure 2.28, curve 2). Usually, the second term in the series (2.38) proves to be sufficient for the adequate description of the experimental phenomena [15]. In Figure 2.28, a particular form of the potential (2.38) is also shown (curve 3). In any case, once the form $W(\theta)$ is chosen, the problem of the search for the director field in a layer of a finite thickness can be solved at the macroscopic level by solving the variation problem of the minimization of a total value of the liquid crystal bulk and surface energy [1,2].

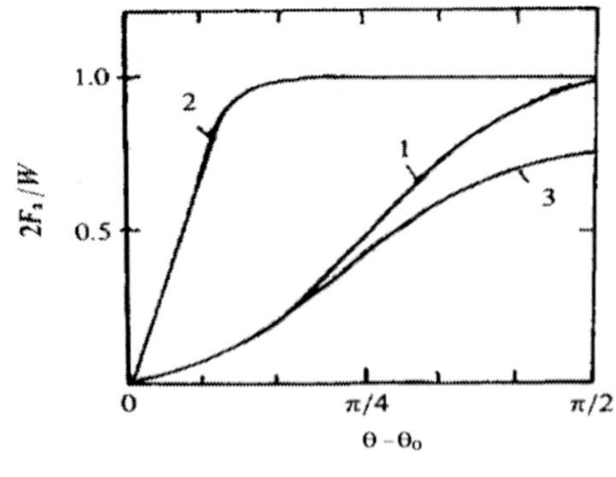

Figure 2.28. Surface anchoring potential as a function of the deviation angle (from the preferred orientation at the surface: (l) "Rapini" potential (2.37), (2) the elliptic sine potential (2.39), (3) Legendre expansion (2.38) with the two first terms [15].

REFERENCES

[1] Chigrinov, VG. Liquid Crystal Devices: Physics and Applications, (*Artech-House*, 1999).

[2] Blinov, LM; Chigrinov, VG. Electrooptical effects in liquid crystalline materials, (*Springer-Verlag*, 1994).

[3] Pasechnik, SV; Chigrinov, VG; Shmeliova, DV. Liquid Crystals, Viscous and Elastic Properties, (*Wiley-VCH*, Berlin, 2009).

[4] de Jeu, WH. Physical properties of liquid crystalline materials, (*Gordon and Breach*, 1980).

[5] Berreman, DW. Solid Surface Shape and the Alignment of an Adjacent Nematic LC, *Phys. Rev. Lett.*, 28, 1683 (1972).

[6] Fukuda, J; Yoneya, M; Yokoyama, H. Surface-Groove-Induced Azimuthal Anchoring of a Nematic Liquid Crystal: Berreman's Model Reexamined, *Phys. Rev. Lett.*, 98, 187803 (2007).

[7] Uchida, T; Seki, H. Surface Alignment of Liquid Crystals, in Liquid Crystal Applications and Uses, ed. By B. Bahadur, (Singapore, *World Scientific*, 1991).

[8] Chigrinov, VG; Kozenkov, VM; Kwok, HS. Photoalignment of liquid crystalline materials: Physics and Applications, (*Wiley*, 2008).

[9] Myrvold, B; Komdo, K. The relationship between chemical structure of nematic liquid crystals and teir pretilt angles, *Liq. Cryst.*, 18, 271 (1995).

[10] Kwok, HS; Yeung, FSY. Nano- structured liquid crystal alignment layers, *J. SID*, 16, 911, (2008).

[11] Jacob, YL. Ho, Chigrinov, VG; Kwok, HS. Variable liquid crystal pretilt angles generated by photoalignment of a mixed polyimide alignment layer, *Appl. Phys. Lett.*, 90, 243506 (2007).

[12] Hyun, SY; Moon, JH; Shin, DM. Application of photoalignment technologies to antiferroelectric liquid crystal cell, *Mol. Cryst. Liq. Cryst.*, 412, 369 (2004).

[13] Barbero, G; Madhusudana, NV; Durand, G. Anchoring Energy of Nematic Liquid Crystals an Analysis of the Proposed Forms, *Z. Naturforssch.*, 39A, 1066 (1984).

[14] Barnik, MI; Blinov, LM; Korkishko, TV; Umanskiy, BA; Chigrinov, VG. Investigation of NLC director orientational deformations in electric field for different boundary conditions, *Mol. Cryst. Liq. Cryst.*, 99, 53 (1983).

[15] Yokoyama, H; van Sprang, HA. A novel method for determining the anchoring energy function at a nematic liquid crystal wall from director distortions at high fields, *J. Appl. Phys.*, 57, 4520 (1985).

Chapter 3

ELECTROOPTICAL EFFECTS IN LIQUID CRYSTAL MATERIALS

Based on the director distribution we can derive the electrooptical response of liquid crystal cell (such as birefringence), rotation of the polarization plane of the incident light, total internal reflection, absorption, or some other important characteristics of the cell. Basic electrooptical liquid crystal modes will be discussed below.

3.1. ELECTRICALLY CONTROLLED BIREFRINGENCE

3.1.1. Static Director Distribution

The most important geometries of Electrically Controlled Birefringence (ECB) are shown in Figure 3.1.

A compromise between dielectric and elastic torques results in director reorientation from the initial alignment $\theta(z)$ with the maximum deviation θ_m at the center of the layer (the Freedericksz transition). The effect occurs when the electric field exceeds a certain threshold value:

$$U_F = \pi(4\pi K_{ii} / \Delta\varepsilon)^{1/2} \tag{3.1}$$

where the LC elastic constant $K_{ii} = K_{11}$ or K_{33} for the splay (S) and bend (B) Freederiksz transitions, respectively (Figures 3.1 (a, b)), and $\Delta\varepsilon$ - LC dielectric anisotropy. The initial director alignment is homogeneous ($\mathbf{E} \perp \mathbf{Lo}$) for the S Freederiksz transition (S-effect) and homeotropic ($\mathbf{E} \parallel \mathbf{Lo}$) for the B-effect. The action of the electric field on the liquid crystal (LC) layer results in the deformation of the initial molecular (director) distribution and a subsequent variation of the LC cell optical properties [1-3]. Director **L** reorients in an electric field under the action of the dielectric torque, which is proportional to the dielectric anisotropy $\Delta\varepsilon$ (2.9). The corresponding contribution g_ε to the density of the nematic free energy gives [1-3]:

$$g_\varepsilon = -\mathbf{DE}/8\pi = -\varepsilon_\perp E^2/8\pi - \Delta\varepsilon\,(\mathbf{EL})^2/8\pi \tag{3.2}$$

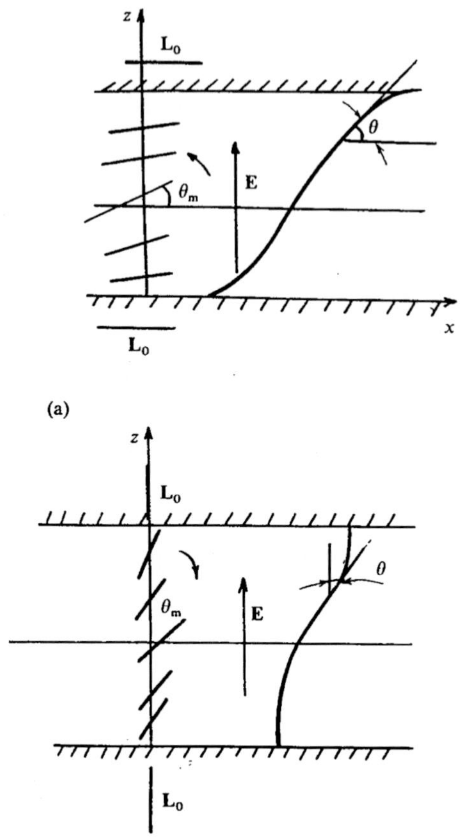

(b)

Figure 3.1. Electrically Controlled Birefringence (ECB) mode for various LC configurations: (a) S-effect (homogeneous alignment, electric field is vertical or perpendicular to the substrates); (b) B-effect or VAN-effect (homeotropic alignment) [1-3].

i.e., director **L** tends to align itself along the field (**L**||**E**) if $\Delta\varepsilon > 0$ and is perpendicular to it, (**L** ⊥ **E**), provided that $\Delta\varepsilon < 0$.

The elastic torque supports the initial director orientation, fixed by the boundary conditions on the surface (S), which in case of a strong surface anchoring writes:

$$\mathbf{L}\big|_S = \mathbf{L}_o \tag{3.3}$$

As a result of this a compromised director profile appears, which satisfies to the condition of the minimum free energy:

$$F_V = \int_V (g_k + g_\varepsilon) d\tau, \tag{3.4}$$

where $g_k = 1/2[K_{11}\,(\mathrm{div}\,\mathbf{L})^2 + K_{22}\,(\mathbf{L}\,\mathrm{rot}\,\mathbf{L} - q_0)^2 + K_{33}\,(\mathbf{L} \times \mathrm{rot}\,\mathbf{L})^2]$ (see also (2.15)) is the elastic energy density and K_{ii} are the elastic moduli ; $q_o = 2\pi/R_o$ characterizes the "natural chirality" and equals to zero in pure nematics. R_0 is the helical pitch in the LC mixture induced by a chiral dopant, usually the value of the helix pitch is inversely proportional to the concentration of the dopant [1-3].

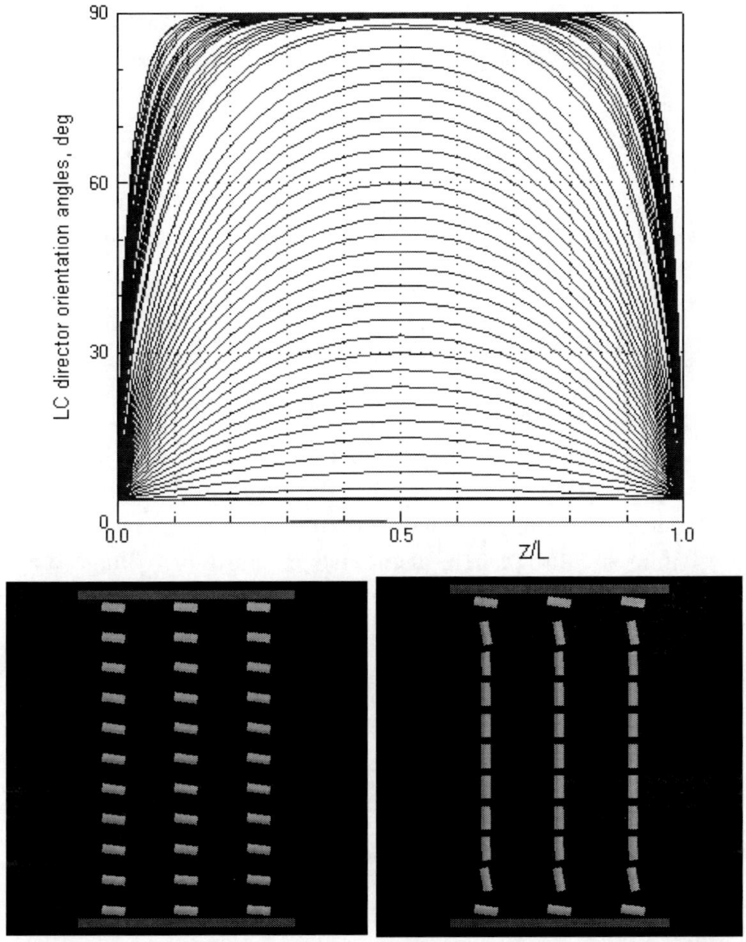

Figure 3.2. Above: LC director distributions for the relative thickness of the cell: 0≤z/d≤1, below: LC molecules arrangement inside LC cell.

In a more general case of a finite director anchoring at the boundaries (see also 2.4.3), we have to write the total energy F of liquid crystals as follows:

$$F = Fv + Fs, \tag{3.5}$$

where F_s is the surface energy, characterized by different anchoring terms, known as "polar" and "azimuthal" anchoring energies [1-3]. The corresponding contributions (2.37) to the anchoring F_s on the surface S have been given above (Figure 2.16).

Typical pictures of LC director distributions for ECB effect are shown in Figure 3.2. Based on the director distribution we can derive the electrooptical response of a nematic liquid crystal cell (such as birefringence), rotation of the polarization plane of the incident light, total internal reflection, absorption, or some other important characteristics of the cell.

3.1.2. Effect of a Weak Anchoring at the Boundaries

It should again be emphasized that, in contrast to the case of a tilted orientation where there is no threshold, the threshold for the Freedericksz transitions (3.1) remains for a weak director boundary anchoring, but has a lower value. The corresponding equation is [1-3]:

$$\text{cotan}[\pi U_F(W)/U_F(oo)] = \pi K_{ii} U_F(W)/WdU_F(oc)), \tag{3.6}$$

where the effective elastic coefficient K_{ii} and the Freederiksz transition threshold U_F are defined for finite (W) and infinite (∞) anchoring energies. For large anchoring energies the expression (3.6) reduces to:

$$U_F\,(W) = U_F\,(\infty)\,(1 - 2K_{ii}\,/Wd),\ Wd/K_{ii} >> 1. \tag{3.7}$$

The value of the effective anchoring energy W depends on the form of the potential barrier W(θ). For sufficiently small values of anchoring energy the electrooptical response of the nematic cell becomes infinitely steep, so that for a certain critical value of W a hysteresis and first-order Freederiks transition becomes possible [1-3].

In the case of finite anchoring there also exists a saturation voltage for the total director reorientation, parallel (Figure 3.1(a)) or perpendicular (Figure 3.1(b)) to the field, when the boundary regions disappear [1-3]. Development of the nematic cells with a good and reliable control of anchoring energy is of great importance for applications, as very steep transmission-voltage curves, memory states, and improved response times could be realized [1-3].

3.1.3. Dynamics of the Director Motion. Back-Flow Effect

It is easiest to examine the kinetics of the director motion for a pure twist or T deformation, which is not accompanied by a change in position of the centers of gravity of the molecules. In this case the electric or magnetic field is applied parallel to the substrates, exerting a torque, which causes a pure twist deformation of the initially homogeneous director alignment. In contrast, for S- and B-Freedericksz transitions the rotation of the director is accompanied by such a change in position, i.e., by a movement of the liquid (back-flow). In order to allow for this back-flow effect the equation of motion of the director is coupled with that of the fluid.

For a pure T deformation the equation of motion of the director expresses the balance between the torques due to the elastic and viscous forces and the external field (and does not contain the fluid velocity) [1-3]

$$K_{22}\partial^2\varphi\,/\partial z^2 + \frac{\Delta\varepsilon E^2}{4\pi}\sin\varphi\cos\varphi = \gamma_1\partial\varphi\,/\partial t \tag{3.8}$$

This equation describes the director rotation in a electric field E with the inertia term $I\partial^2\varphi/\partial t^2$, which is omitted, $\gamma_1 = \alpha_3 - \alpha_2$ is a rotational viscosity, and α_i are Leslie viscosity coefficients. Equation (3.8) in the limit of small φ angles transforms to:

$$K_{22}\partial^2\varphi/\partial z^2 + \frac{\Delta\varepsilon E^2}{4\pi}\varphi = \gamma_1\,\partial\varphi/\partial t \tag{3.9}$$

with the solution $\varphi = \varphi_m \exp(t/\tau_r)\sin(\pi z/d)$, where

$$\tau_r = \gamma_1/(\frac{\Delta\varepsilon E^2}{4\pi} - K_{22}\,\pi^2/d^2) \tag{3.10}$$

is the reaction or switching on time. The solution (3.10) satisfies the strong anchoring boundary conditions $\varphi(z=0) = \varphi(z=d) = \varphi_m$ and assumes a maximum value at the center of the layer $\varphi(z=d/2)=\varphi_m$.

The corresponding relaxation or decay times are found from (3.9) for E = 0 in a similar way:

$$\tau_d = \gamma_1\, d^2/K_{22}\pi^2 \tag{3.11}$$

Unlike T-deformation, the director reorientation in S- and B- effects is always accompanied by a macroscopic flow of a nematic liquid crystal (back-flow) with the velocity **V** = (V(z), 0, 0), where the z-axis goes perpendicular to the substrates and the deformations take place in the x, z-plane. The velocity V includes only the x-component, because the z-component is zero according to the continuity equation (div **V** = 0), and the y-component vanishes due to the symmetry of the problem.

For small variations of the angle θ the characteristic times of the S and B-effect can be found from the solutions of the coupled dynamic equations of the nematic director θ (z) and velocity V(z) in the following form:

$$\tau_r = \gamma_1^*/(\Delta\varepsilon\, E^2/4\pi - K_{ii}\pi^2/d^2), \qquad \tau_d = \gamma_1^*/K_{ii}\pi^2/d^2 \tag{3.12}$$

where γ^*_1 is the effective rotational viscosity. We have:

$$\begin{aligned}\gamma_1^* &= \gamma_1 - 2\alpha_3^2/(\alpha_3+\alpha_4+\alpha_6),\ \ K_{ii} = K_{11}\ \text{for S-effect}\\ \gamma_1^* &= \gamma_1 - 2\alpha_2^2/(\alpha_4+\alpha_5-\alpha_2),\ \ K_{ii} = K_{33}\ \ \text{for B-effect}\end{aligned} \tag{3.13}$$

The relationships (3.13) show that back-flow considerably alters the response times and should be taken into account.

The relative difference $(\gamma_1 - \gamma_1^*)/\gamma_1$ becomes close to zero for voltages slightly above the threshold in the S-effect and very high voltages in the B-effect and can hardly exceed 50% within the whole voltage interval. The results of computer simulation were confirmed in experiment [1-3].

Weak boundary anchoring decreases a rise time τ_r time and increases the relaxation time τ_d. It can be easily understood, if we make the substitution $K_{ii}\pi^2/d^2 \Rightarrow 2W/d$ in formulas (3.10)-(3.11), which proves to be correct for sufficiently low anchoring energies [1-3].

3.1.4. Optical Response

3.1.4.1. Splay-mode (S-effect)

To understand the optical characteristics of a liquid crystal layer in electrically controlled birefringence (ECB) effect let us consider the geometry of Figure 3.1, a with the initial homogeneous director orientation along x-axis. If the applied voltage is below the threshold, the nematic liquid crystal layers manifest birefringence $\Delta n = n_e - n_o = n_{\|} - n_{\perp}$. When the field exceeds its threshold value, the director deviates from its orientation along the x-axis, while remaining perpendicular to the y-axis in xy plane. Therefore, the refractive index for the ordinary ray remains unchanged, $n_o = n_{\perp}$. At the same time, the refractive index for the extraordinary ray (n_e) decreases, tending towards n_o. It is simple to relate the magnitude of n_e, to the angle of orientation of the director $\theta(z)$ [1-3]:

$$n_e(z) = n_e n_o / \sqrt{n_e^2 \sin^2\theta(z) + n_o^2 \cos^2\theta(z)} \qquad (3.14)$$

The phase difference between the extraordinary and the ordinary ray for a monochromatic light of wavelength λ is found by integrating over the layer depth (d):

$$\Delta\Phi = 2\pi/\lambda \int_0^d (n_e(z) - n_0)dz = 2\pi\, d < \Delta n(z) > /\lambda \qquad (3.15)$$

The intensity of the light passing through the cell depends on the angle φ_0 between the polarization vector of the incident beam and the initial orientation of the director of the nematic liquid crystal

$$I = I_0 \sin^2 2\varphi_0 \sin^2(\Delta\Phi/2) \qquad (3.16)$$

where I_o is the intensity of the plane polarized light incident on the cell. Hence, the external magnetic or electric field changes the orientation of the director, $\theta = \theta$ (E, z) and, consequently, the values of Δn (E, z) and $\Delta\Phi(E)$. A change in the phase difference $\Delta\Phi$, in turn, results in an oscillatory dependence of the optical signal at the exit of the analyzer. The maximum amplitude of these oscillations corresponds to an angle $\varphi_0 = 45^0$ and the maximum possible number of oscillations (e.g., the number of maxima during a complete reorientation of the director) is approximately $(n_{\|} - n_{\perp})d/\lambda$.

The characteristic curves of transmitted intensity in the ECB effect are shown in Figure 3.3. The contrast ratio I_{max}/I_{min} is not, in principle, limited (3.16), but determined in experiment by the quality of the initial director orientation. The output light, in general, becomes elliptically polarized, so that its ellipticity e and the angle ψ, between the long ellipse axis and the polarizer, also depend on φ_0 and $\Delta\Phi$ [1-3]:

$$\begin{aligned} e &= \tan[1/2\arcsin(\sin 2\varphi_0 \sin \Delta\Phi)] \\ \tan 2\psi &= \tan 2\varphi_0 \cos \Delta\Phi \end{aligned} \tag{3.17}$$

The experimental dependences of e(U) and ψ(U) on the applied voltage resemble those shown in Figure 3.3 for the intensity curve I(U) [1-3].

The effect of phase modulation for an initial planar orientation of the director (along x) with positive dielectric anisotropy ($\Delta\varepsilon>0$) and with the field applied along the z-axis (Figure 3.1(a)) is called the S-effect [1-3], since the initial deformation is a splay deformation, even though a bend deformation is also induced above the threshold.

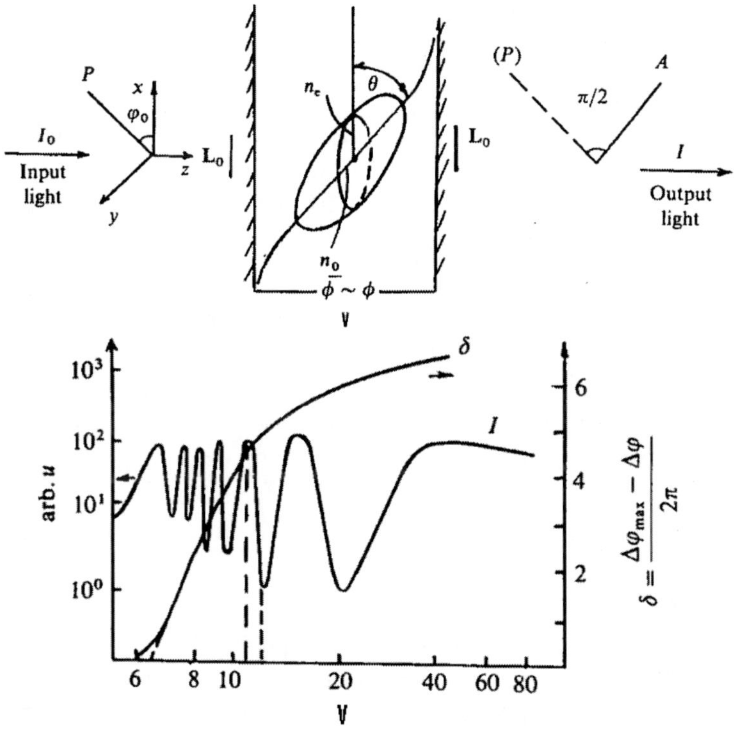

Figure 3.3. Electrically controlled birefringence in homogeneous liquid crystal cell (S-effect). Polarizer (P) is crossed with the analyzer (A) and set at the angle φ_0 with respect to the initial liquid crystal director. Below the intensity I of the transmitted light and phase difference δ are shown versus the applied voltage. The phase π-switching regime is indicated, which results in the variation of the intensity from I_{max} to I_{min} [1-3].

The applied field causes a phase difference to arise between the ordinary and extraordinary rays and the intensity of the light oscillates in accordance with (3.16). In case the bend deformation is now in the initial stages of its development (Figure 3.1,b), the corresponding electrooptical effect is called B-effect, taking place for negative values of a dielectric anisotropy $\Delta\varepsilon<0$. However, here the final orientation of the director is not defined (degenerate), and the sample does not remain monodomain and contains many specific defects (umbifics) [1-3]. In principle, the preferred direction of the final orientation of the director can be established, and these defects in the structure are eliminated by a special preparation of the surfaces with a slight pretilt.

The important problem of the ECB effect is to provide full switching from I_{max} to I_{min} with minimal response times. According to (3.16), such a switching is stipulated when the phase difference $\Delta\Phi$ is changed as much as by π. If a nematic cell is subjected to voltage U corresponding to a maximum intensity I_{max}, then to attain another state with I_{min} we have to supply an additional voltage $U_\pi \geq \Delta U$, where ΔU is the minimum possible value of U_π, (Figure 3.3).

The main disadvantage of the S-effect for display applications is the strong dependence of the transmitted intensity on the light wavelength, and the nonuniform transmission-voltage characteristics at oblique light incidence. It is possible, however, to avoid them by placing a compensating birefringent plate between the liquid crystal cell and one of the polarizers or by using two nematic S-cells in series which have perpendicular initial directors [1-3].

3.1.4.2. π-cells

Special attention should be paid to the so-called π-cells, when the birefringent intensity is switched in the last fall of the oscillation curve (Figure 3.3) [1-3]. In this case the switching is attained due to the very slight variation of the director distribution within the narrow regions near the boundaries, thus resulting in a very fast response speed. The corresponding switching times can be estimated according to the formula [1-3]:

$$\tau \approx (\Delta/\pi/2\pi)^2\, 1/(1-\beta U_\pi/U_0)^2\, \gamma_1\lambda^2/K_{11}\Delta n^2, \tag{3.18}$$

where $\Delta/\pi \approx 1$ is a relative phase difference for the last intensity fall, $\beta \approx 1$-the liquid crystal material constant. As we may see, the response time does not depend on the cell thickness. Combining π cells with phase retardation plates, both with a positive and negative phase shifts it is possible to optimize a contrast and color uniformity of the liquid crystal cell [1-3].

3.1.4.3. Bend-mode (B-effect)

B-effect in homeotropic or quasi-homeotropic (slightly tilted) nematic samples remains attractive for applications, including displays with high information content [1-3]. With good homeotropic orientation of a nematic liquid crystal, B-effect is characterized by a steep growth in the optical transmission with voltage, i.e., the threshold is very sharp. This is due to the very weak light scattering of the homeotropically oriented layer, and the complete absence of birefringence in the initial state with crossed polarizers [1-3]. Basically, the patterns observed experimentally are similar for the S- effect, including dynamic behavior. Let us consider the transmission-voltage curves I(U), for the B-effect at voltages slightly exceeding the threshold value, which are important for applications. As mentioned above, to avoid degeneracy in the director reorientation a slight initial pretilt from the normal to the substrate is needed $\theta_0 \approx 0.5\text{-}1^0$ (Figure 3.1,b). The electrooptical response is very sensitive to the θ_0 value, i.e., small $\theta_0 < 0.5^0$ do not allow us to avoid defects, while larger angles, $\theta_0 > 1^0$, strongly reduce the contrast [1-3]. For the voltages slightly exceeding the threshold, the contrast is [1-3]:

$$C = I_{on} / I_{off} \approx \sin^2(A < \theta_{on}^2 >) / \sin^2(A\theta_{off}^2) \approx \theta_m^4 / 4\theta_0^2 \tag{3.19}$$

where $< \theta_{on}^2 >= d^{-1} \int_0^d \theta^2(z)dz \approx \theta_m^2 / 2$ for small θ and $\theta_{\text{off}} = \theta_0$

are the director angles in the switched-on and switched-off states, $A=\pi d n_\perp(1- n_\perp^2/n_\parallel^2)/2\lambda$ - a phase factor. The contrast ratio C crucially depends on the pretilt angle θ_0. We can obtain the following relation for Θ_m^2 in the B-effect:

$$\Theta_m^2 = 4\,(U/U_B - 1)\,(K_{11}/K_{33} + \varepsilon_\parallel / \varepsilon_\perp)^{-1} \tag{3.20}$$

where $U_B = \pi\,(4\pi\,K_{33}/|\Delta\varepsilon|)^{1/2}$ is the B-effect threshold voltage. Using the expression (3.16) for $\varphi_0 = \pi/4$ we derive the following relationship for the optical transmission $T = I/I_0$ in the ECB effect in a homeotropic nematic (B-effect):

$$T=\sin^2\pi d\Delta n\lambda^{-1} <\theta_{on}>^2 \approx 4A^2\,(U/U_B - 1)^2\,(K_{11}/K_{33} + \varepsilon_\perp/\varepsilon_\parallel - 1)^{-2} \approx [2\pi d\lambda^{-1}\,(n_\parallel - n_\perp)/\,(K_{11}/K_{33} + \varepsilon_\perp/\varepsilon_\parallel - 1)]^2\,(U/U_B - 1)^2 \tag{3.21}$$

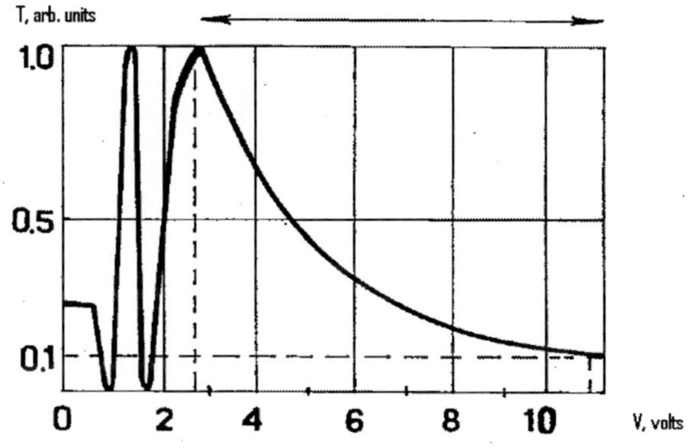

Figure 3.4. The regime of switching in π-cell . For the last intensity fall the controlling voltage has to change from U_π to U_0[1-3].

As seen from (3.21), a steep electrooptical response of the cell is attained for sufficiently large values of the optical path difference $d(n_\parallel - n_\perp)$ and the elasticity anisotropy K_{33}/K_{11}, as well as for small dielectric anisotropy $|\Delta\varepsilon|/\varepsilon_\parallel$. However, the values of d can not be too large, since the latter results in an increase of response times (3.11), while small values of the dielectric anisotropy lead to a growth in operating voltages. Thus, to develop a good ECB LC material, a compromise is needed. In order to obtain the steep electrooptical characteristics required for displays with high-information content, together with a fast response and uniformity of transmission for the oblique light incidence, the following parameters of nematic cells for the B-effect can be proposed [1-3]: small thickness, d < 5 μm; small pretilt angle $\theta_0 \approx 0.5$-1^0; large optical retardation $d\Delta n \approx 1$ micron; large K_{33}/K_{11} ratio; and small dielectric anisotropy $|\Delta\varepsilon|/\varepsilon_\parallel < 0.5$. The regime of π-cell switching is shown in Figure 3.4.

3.1.4.4. Hybrid Aligned (HAN) Nematic Mode

Together with π-cell, this is interesting to consider a hybrid aligned (HAN) nematic mode (Figure 3.5). In this mode the director is homogeneous on one LC substrate and homeotropic on the other (Figure 3.5), so the combination of S and B deformations takes place as a result. HAN mode has no threshold and possess the smooth electrooptical characteritics with a number of gray levels (Figure 3.5). The geometry of the mode is similar to S-effect (Figure 3.1a). The maximum possible transmission is obtained, when the structure is placed between crossed polarizers and the polarizer angle is 45^0 with respect to the plane of SB deformation (Figure 3.5).

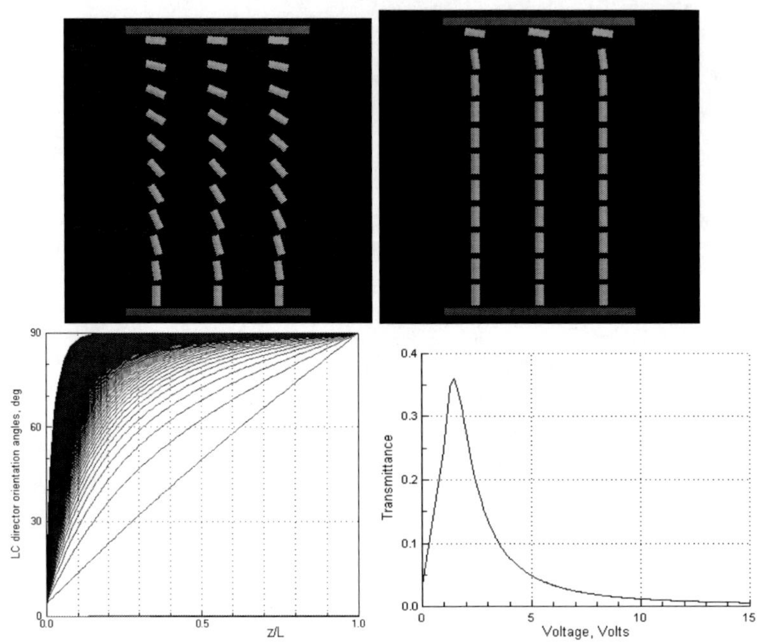

Figure 3.5. HAN LC configuration (above) and its electrooptical response (below), including the director orientation angles (below left) and transmittance in crossed polarizers (below right).

3.2. Twist Effect

If the directions x and y of the planar orientation of nematic liquid crystal molecules on opposite electrodes are perpendicular to each other and the material has a positive dielectric anisotropy $\Delta\varepsilon > 0$, then when an electric field is applied along the z-axis (Figure 3.6) a reorientation effect occurs which is a combination of the S, B, and T deformation [1-3].

In the absence of the field, the light polarization vector follows the director and, consequently, the structure rotates the polarization plane up to the angle characterizing the structure $\varphi_m = \pi/2$ (Figure 3.6). This specific wave guide regime (the Mauguin regime) takes place when

$$\Delta nd/\lambda >> 1 \qquad (3.22)$$

When the applied voltage exceeds a certain threshold value

$$U_{tw} = \pi[\pi\,(4K_{11} + K_{33} - 2K_{22})/\,\Delta\varepsilon\,]^{1/2} \tag{3.23}$$

the director **L** deviates from the initial orientation so that the linear dependence of the azimuthal angle φ(z) disappears and the tilt angle θ(z) becomes nonzero. The qualitative character of the functions φ(z) and *θ(z)* for different voltages is shown in Figure 3.7.

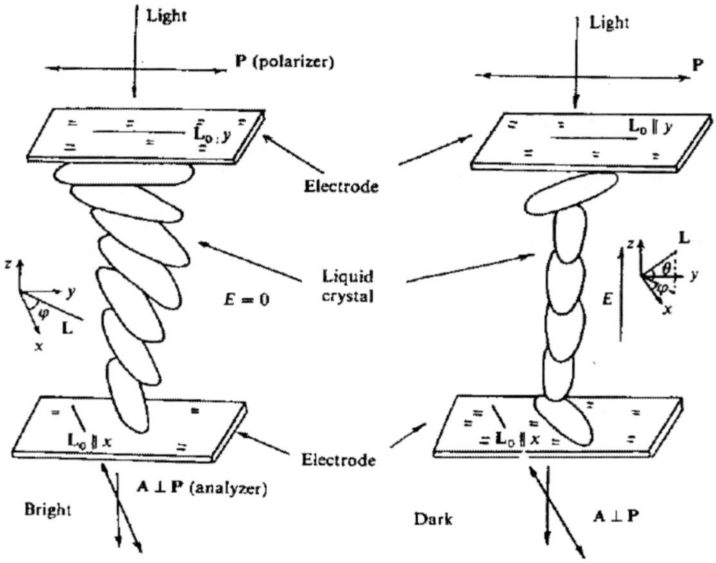

Figure 3.6. Twist effect.

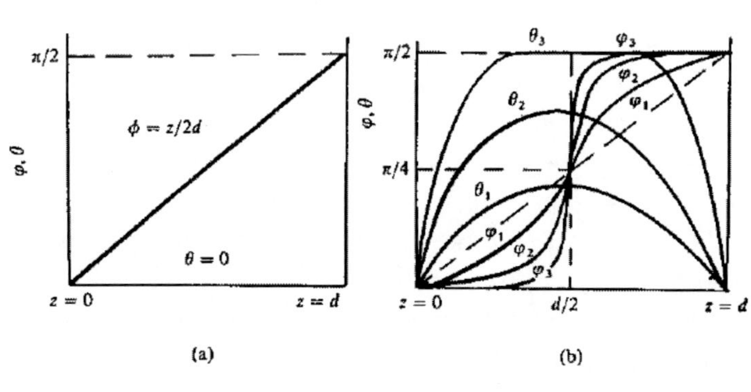

Figure 3.7. Distribution of the director angles φ(z) and θ(z) in twist effect for different voltages: (a) $U \leq U_{tw}$, (b) $U_{tw} < U_1 < U_2 < U_3$.

Since the director tends to orient perpendicular to the substrates, the effective values of Δn and d decrease and, for a certain voltage (optical threshold of the twist U_{opt}), the waveguide regime no longer remains. Let us note that, despite the fact that the director starts to reorient at $U = U_{tw}$, a visible change of the twist-cell transmission is observed only for $U = U_{opt} > U_{tw}$ (Figure 3.8).

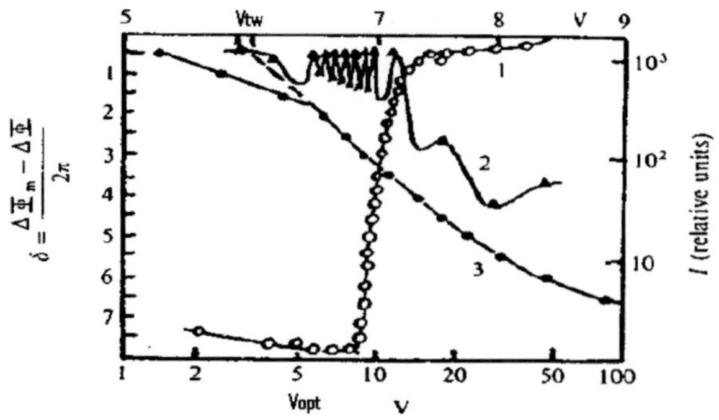

Figure 3.8. Optical response of the twist-cell between parallel polarizers [1-3]. Curve 1: polarizers are parallel to the director on the input surface of the cell (conventional orientation); curve 2: polarizers at the angle of 45° to the director on the input surface (maximum birefringence intensity); curve 3: phase retardation (3.15) calculated from curve 2 [1-3].

Figure 3.8 gives the dependence of the optical transmission of a twist cell for both the conventional geometry **P**|| **L**(0) and when the polarizer **P** forms an angle of 45° with respect to the orientation of the director at z = 0 [1-3]. The deformation threshold, U_{tw} = 6 V, determined by extrapolating the linear section of the phase delay curve to δ(U) = 0, coincides with that calculated from (3.23). The optical threshold for the twist effect increases with decreasing the wavelength (U_{opt} = 8.9 V and 10.2 V for λ = 750 and 450 nm, respectively), since the cut-off implied by the Mauguin condition occurs at higher voltages for shorter wavelengths (3.22).

3.2.1. Twist-Cell Geometry for Zero Voltage. Mauguin Conditions

A twist cell is usually formed by placing orienting glasses on top of each other. Then twist directions at angles π/2 and -π/2 are equally probable. This degeneracy in the sign of the twist can be removed if small amounts of optically active material is added to the nematic liquid crystal. In this case, the walls disappear and the twist cell has a uniform structure throughout its entire area. A nonzero tilt of the molecules to the cell surface θ_0 also results in a non-uniform twisting. At zero field the waveguide regime or Mauguin limit is violated for small values of Δnd/λ. Figure 3.9 shows the corresponding dependence of the transmitted light intensity (I), ellipticity (e), and the rotation angle (ψ) of the long ellipse axis with respect to the polarizer P-axis on the parameter Δnd/λ [1-3] (the director at the first substrate is parallel (or perpendicular) to the polarizer (P) and analyzer (A), (**P** || **A**)). As seen from Figure 3.9 the exact Mauguin conditions

$$I=0,\ \psi=\pi/2,\ e=0 \tag{3.24}$$

take place not only for infinitely large Δnd/λ values, but also at some discrete points

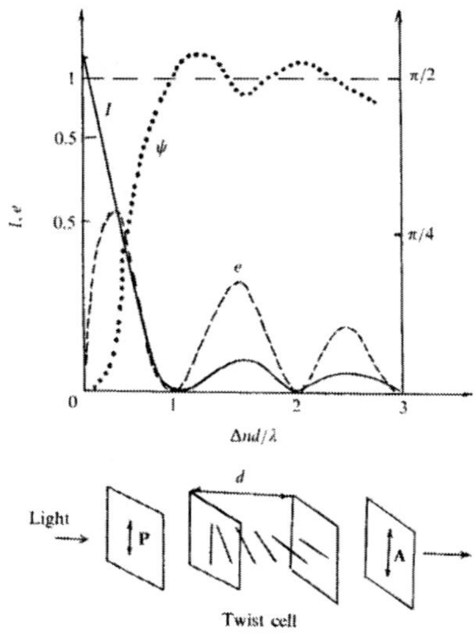

Figure 3.9. Optical characteristics of the twist-cell in the absence of the field. I, intensity (solid line), e, ellipticity (dashed line), θ, rotation angle (dotted line) versus the parameter Δnd/λ. Without the twist cell (d = 0), the intensity between the parallel polarizer (P) and the analyzer (A) is taken equal to I = 1. The director at the first substrate of the twist cell is parallel to the polarizer (P) [1-3].

$$\Delta nd/\lambda = (4m^2-1)^{1/2}/2, \quad m=1,2,3, \ldots \tag{3.25}$$

usually called Mauguin minima.

We may consider the transmission of the twist cell for a white light, thus eliminating λ from the characteristic dependence of the twist cell transmission. Writing T(Δnd/λ) it is possible to average T(λ), together with the function of the sensitivity of the human eye Y(λ), and the wavelength distribution of the illumination source H(λ) [1-3]. The corresponding optimal points, which provide the minimum transmission of the twist cell between parallel polaroids, are very close to those defined by (3.25), if we take the wavelength of the maximum sensitivity of the human eye in the range λ≈ 550-580 nm.

3.2.2. Transmission-Voltage Curve for Normal Light Incidence

The typical Transmission-Voltage curve (TVC) of the twist effect for normal light incidence is shown in Figure 3.10 for the twist cell placed between parallel polarizers. As mentioned above, the deformation of the director, measured by birefringence or capacity, begins for smaller voltages $U = U_{tw}$, then visible variations of the twist cell transmission. In view of the Mauguin requirement (3.22), the optical threshold of the twist effect decreases for the smaller values of the cell thickness and the optical anisotropy of the liquid crystal as well as for director pretilt on the boundaries [1-3]; $\theta = \theta_0$ at z=0,d.

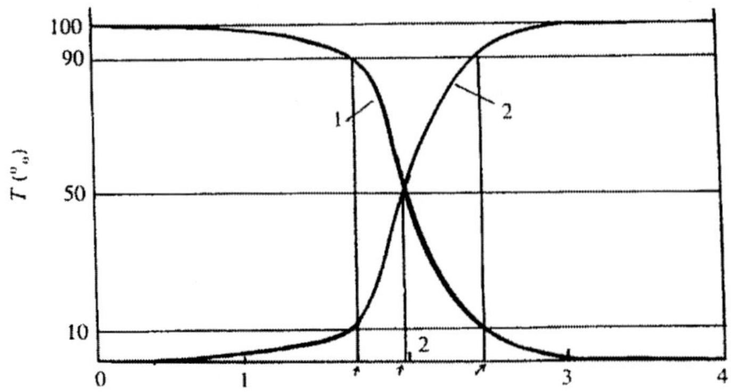

Figure 3.10. Optical transmission of the twist cell versus voltage for crossed (curve 1) and parallel (curve 2) polarizers [1-3]. The voltages U_{90}, U_{50}, and U_{10}correspond to the 90%, 50%, and 10% transmission level of curve 1, respectively.

One of the most important parameters of a twist cell is the steepness of the TVC. Usually, the steepness parameter p is defined from the TVC of the twist cell, placed between crossed polarizers (Figure 3.10),

$$p_{50} = U_{50}/U_{90} - 1,\ p_{10} = U_{10}/U_{90} - 1 \tag{3.26}$$

where U_{90}, U_{50}, and U_{10} correspond to the 90%, 50%, and 10% levels of the optical transmission. As seen from (3.26), the steeper TVCs correspond to the smaller values of p_{50} and p_{10}.

The steepness of the TVC could be optimized for any specific case. There is no need to perform numerous experiments as the computer simulation could solve the problem with a high degree of accuracy. At present, many authors considered the problem of the calculation of electrooptical characteristics of twist cells, using both analytical estimations and computer programs [1-3]. The procedure of the modeling of the twist cell electrooptical behavior includes two steps. First, the distributions of the director orientation are found for varying voltages, which actually characterizes the distribution of the local LC optical axis (Figure 3.7). The second step consists of solving Maxwell equations for the light propagating in anisotropic liquid crystal media. To accomplish this, the liquid crystal is divided into a series of N equally thick uniaxial crystal sublayers, possessing a uniform optical axis direction within each sublayer. The direction of the optical axis changes by a small angle between adjacent sublayers, thus following the total director distribution in the cell (Figure 3.11). Neglecting multiple reflections decouples forward and backward propagating waves reducing the 4 x 4 formalism to a 2 x 2 calculus [4]. The polarization state in a sublayer is determined only by the preceding sublayers. Another way for constructing 2 x 2 matrix methods is from bottom up [5]. At the entrance of each layer the refracted wave is decomposed into two eigenwaves of the layer. With the 2 x 2 propagator matrix the eigenwaves at the exit of the medium are calculated. A further decomposition of the field vectors onto the directions parallel and perpendicular to the plane of incidence gives the two amplitudes of the transmitted light waves. Reflection can be included by using the Fresnel formulas at each interface (Figure 3.11). Each sublayer is represented by a propagation matrix which alters the polarization state of the light passing through it, so that the total twist cell transmission is found by multiplying all the matrices of sublayers. The computer methods for calculating the twist-cell properties were developed by Berreman, using the 4 x 4 propagation matrix [5]. Later on, computer procedures were improved, simplified by neglecting multiple reflections in the sublayers, and made faster. The 2 x 2 matrix formalism, known as “Extended Jones Matrix Method” was developed later [5]. Thus the possibility of using the 2 x 2 Jones matrix which considerably accelerates the procedure without loss of accuracy was demonstrated. At present, several groups [6-9] adjust the set of computer programs as commercially available products for the simulation of electrooptical properties of twist and other nematic cells.

The larger values of the layer thickness lead to the higher operating voltages of the twist effect [1-3]. This is evident, because the Mauguin parameter $\Delta nd/\lambda$ increases in the "off" state and in order to break Mauguin's condition (3.22) the smaller Δn values in the "on" state are needed, corresponding to the stronger director deformation in higher fields. As mentioned above, TVC in parallel and perpendicular polarizers are complementary (Figure 3.10). This is not true, however, for the contrast ratios as functions of applied voltages for a non-monochromatic (white) light. The contrast ratio C is defined as the ratio of transmitted luminances in the on and off-state [1-3]:

$$C=B_{on}/B_{off} \tag{3.27}$$

where

$$B = \int_{\lambda} H(\lambda)\, y(\lambda) I(\lambda) d\lambda \Big/ \int_{\lambda} H(\lambda)\, y(\lambda) d\lambda$$

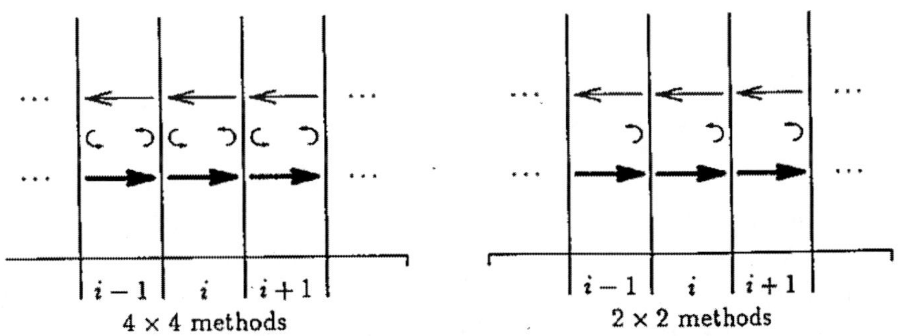

Figure 3.11. Solving of the problem of light propagation in LC anisotropic medium with a special distribution of the optical axis (Figure 3.7). Coupling of forward and backward propagating waves when using 4 x 4 matrix methods and decoupling when using 2 x 2 matrix methods.

The intensities $I(\lambda)$ in the on- and off-states are averaged with the function of the sensitivity of the human eye $y(\lambda)$ and the energy distribution of imination source $H(\lambda)$ over the visible spectrum (380-780 nm). The electrooptic effect in the twist cell placed between parallel and crossed polarizers is called, the "normally black" and "normally white" mode [1-3], in accordance with the appearance of the twist cell in the off state (dark or bright). Contrast ratios in the white mode are considerably higher than in the black mode, as the transmission in the on state for a normally white mode could be very small limited only by the quality of polarizers and liquid crystal orientation.

3.2.3. Twist LCD Electrooptic Response at Various Angles

The twist cell transmission at oblique incidence depends on the values of the polar i_θ and azimuthal i_φ angles of light incidence (Figure 3.12). This could be interpreted in terms of the corresponding Mauguin parameter $\Delta nd/\lambda$, which becomes a function of the light direction. The Mauguin parameter in the direction **e** is estimated by averaging the value of $p\Delta n/\lambda$ along **e**, where $p = (d\varphi/de)^{-1}$ is a local value of the pitch, i.e., the distance of the total director azimuthal rotation by 2π (for **e** parallel to the twist axis p = 4d) and the effective optical anisotropy $\Delta n = (\sin^2\theta/n_{\|}^2 + \cos^2\theta/n_{\perp}^2)^{-1/2} - n_{\perp}$ with the polar θ and azimuthal φ angles of the director with respect to the **e** axis.

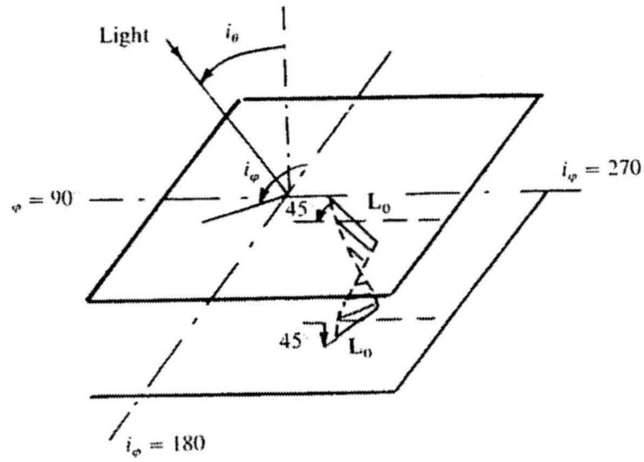

Figure 3.12. The geometry of the twist cell for an oblique light incidence. The angle i_θ is a polar angle of light incidence and the angle i_φ is the corresponding azimuthal angle measured from the symmetric plane of the twist structure [1-3].

The characteristics of transmission for oblique incidence $i_\theta \neq 0$ can be described in terms of the azimuthal dependence of the transmission $T(i_\varphi)$ for a given polar angle of incidence i_θ and the applied voltage U. For directors parallel to the boundaries the twist-cell transmission is symmetric with respect to the plane located at an angle of 45° to the director orientation on the boundaries [1-3]. However, for nonzero director pretilt angles the symmetry is broken. In the T-V curves, obtained for oblique light incidence there appears a minimum of transmission, which goes to the lower voltages for the higher incidence angles (Figure 3.13). Indeed, for a certain voltage the direction of the light propagation may coincide with the director at the center of the layer, thus providing very low values of the Mauguin parameter, despite the fact that the director distribution, as a whole is far from the homeotropic configuration. As seen from Figure 3.13, a twist cell for the oblique light incidence is the most sensitive to an external voltage for azimuthal angles i_φ= 180°, while i_φ= 0° corresponds to quite an opposite case.

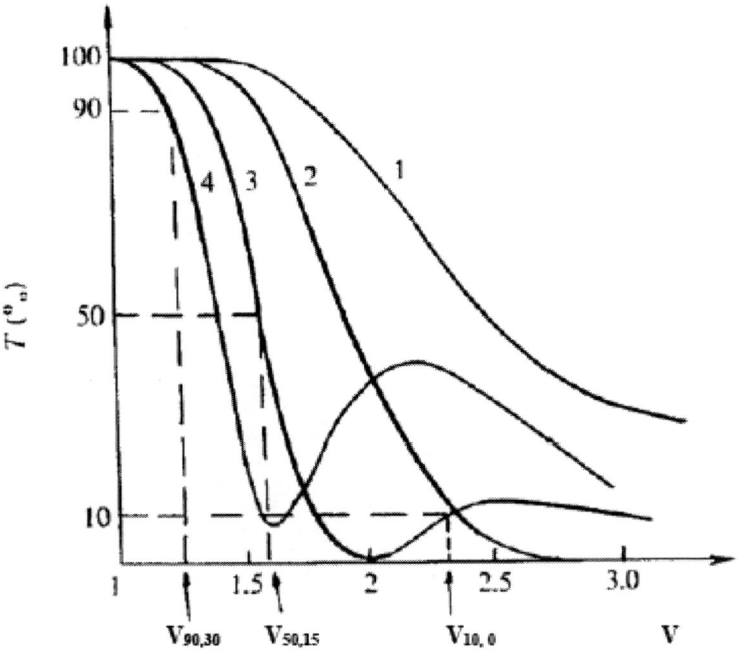

Figure 3.13. TVC of the twist cell for an oblique light incidence [1-3]. The curve 1 corresponds to the light incidence angles $i_\theta=15^0$, $i_\varphi=0^0$; 2- ($i_\theta=0^0$, $i_\varphi=0^0$); 3-($i_\theta=15^0$, $i_\varphi=180^0$); 4-($i_\theta=30^0$, $i_\varphi=180^0$) [1-3].

The most crucial parameter that affects the uniformity of transmission is the Mauguin number m (3.25). For low $\Delta nd/\lambda$ values the anisotropy of transmission for an oblique incidence is weak [1-3]. According to this, liquid crystal mixtures with low Δn values and the first Mauguin minimum $\Delta nd/\lambda = \sqrt{3}/2$ as an operating point are preferable. We do not consider the influence of the parameters K_{33}/K_{11} and $\Delta\varepsilon/\varepsilon_\perp$ on the transmission characteristics of the twist cell, as their effect now is very small compared with the above mentioned case of normal incidence.

Let us note that coloration, i.e., dependence of the light transmission on the wavelength, and the stronger temperature dependence of Δn make the operation in the first Mauguin minimum less attractive. Sometimes it seems more convenient to choose $\sqrt{3}/2 < \Delta nd/\lambda < \sqrt{5}/2$ (between the first and second minima) even at the cost of the partial loss of the contrast.

The dependence of the T-V curves (TVC) on the angles of incidence is taken into account as one of the merit characteristics of twist-effect mixtures. Figure 3.13 shows that for a certain azimuth of an oblique incidence ($i_\varphi=180^0$) the optical threshold of the twist effect is lower than for the normal incidence. Thus, it is possible to consider new definitions of the switching off voltage on a nonselected display element and switching on voltage on the selected one. For instance, Figure 3.13 demonstrates that the optical threshold can be $U_{90,30}$ ($i_\theta=30^0$, $i_\varphi=180^0$, transmission 90%), while the selected voltage can be $U_{50,15}$ ($i_\theta=15^0$, $i_\varphi=180^0$, transmission 50%) or $U_{10,0}$ ($i_\theta=0^0$, transmission 10%). Similar limitations are imposed on the TVC steepness by the temperature dependence of the operating voltages. As a result, the multiplexing capability of the twist effect remains several times lower than that estimated from TVC at normal incidence and room temperature.

For applications it is convenient to evaluate the angular dependence of the transmission by isocontrast curves, which show the levels of equal contrast ratio for different angles of incidence. The examples of these curves for the normally white mode (twist cell between crossed polarizers) and normally black mode (twist cell between parallel polarizers) are given in Figure 3.14. The radial coordinate in the isocontrast diagram defines the value of the incidence angle i_θ, while the azimuthal one defines the azimuthal incidence angle i_φ. As seen from Figure 3.14, the normally black mode provides wider viewing and more uniform viewing angles than normally black mode in twist LC cell.

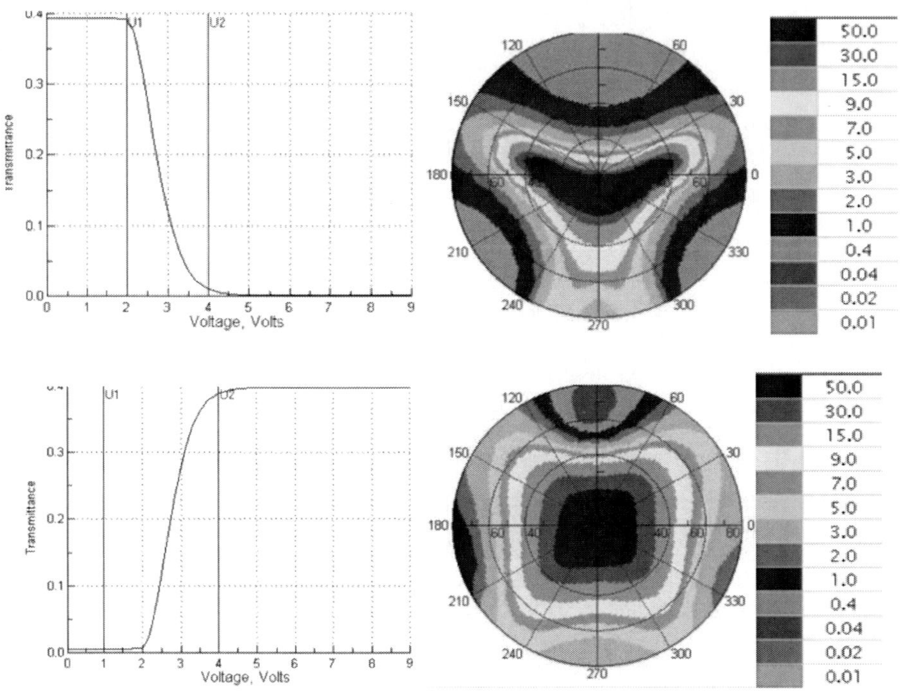

Figure 3.14. Viewing angles of normally white (above) and normally black (below) TN-LCD. The TVC curves of the modes are shown in the left. The viewing angles are defined by contrast ratios at different viewing angles.

The director deformation in TN cell is shown on Figure 3.15. As we have mentioned above to calculate the director deformation angles (Figure 3.7), we need to know all the LC elastic constants K_{11}, K_{22} and K_{33}, dielectric parameters $\varepsilon_{\parallel}$ and $\varepsilon_{\perp}$ and viscosity coefficients α_i . If we are not interested in back flow effect, the information on the rotation viscosity $\gamma_1 = \alpha_3 - \alpha_2$ will be quite sufficient. This is worthwhile to mention, that LC configuration parameters, e.g. LC thickness ratio to helix pitch, LC pretilt angles, and twist angle as well as azimuthal and planar anchoring energies on two LC substrates are also very important parameters, that define LC director distribution (Figure 3.15). In particular, if the anchoring energy on one of the substrates is not high enough TN structure, which still remains even for very high voltages in a strong anchoring case (Figure 3.16), can practically disappear for a low polar anchoring energy on one of the substrates (Figure 3.17).

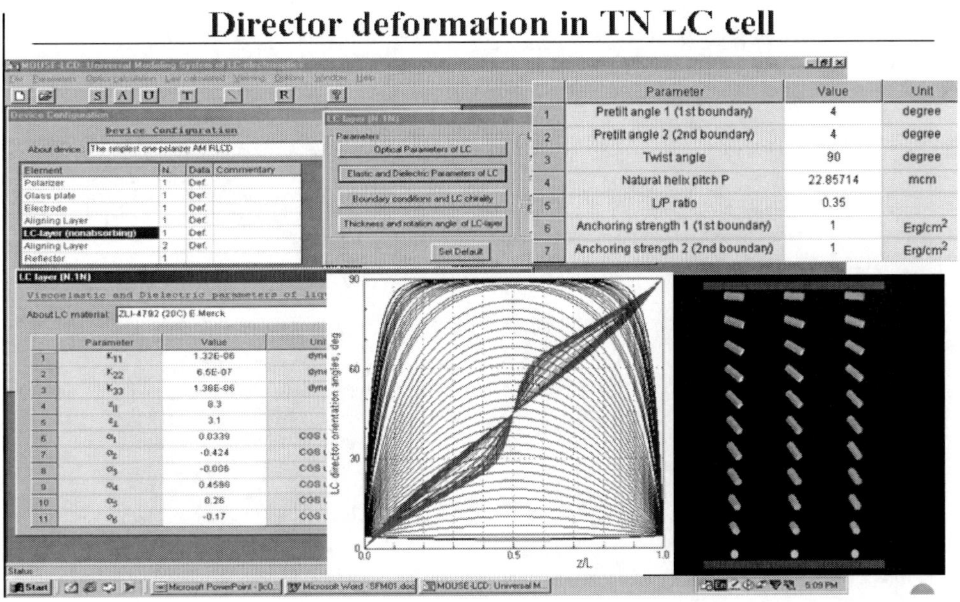

Figure 3.15. Director deformation in TN cell. The interface of MOUSE-LCD program [4].

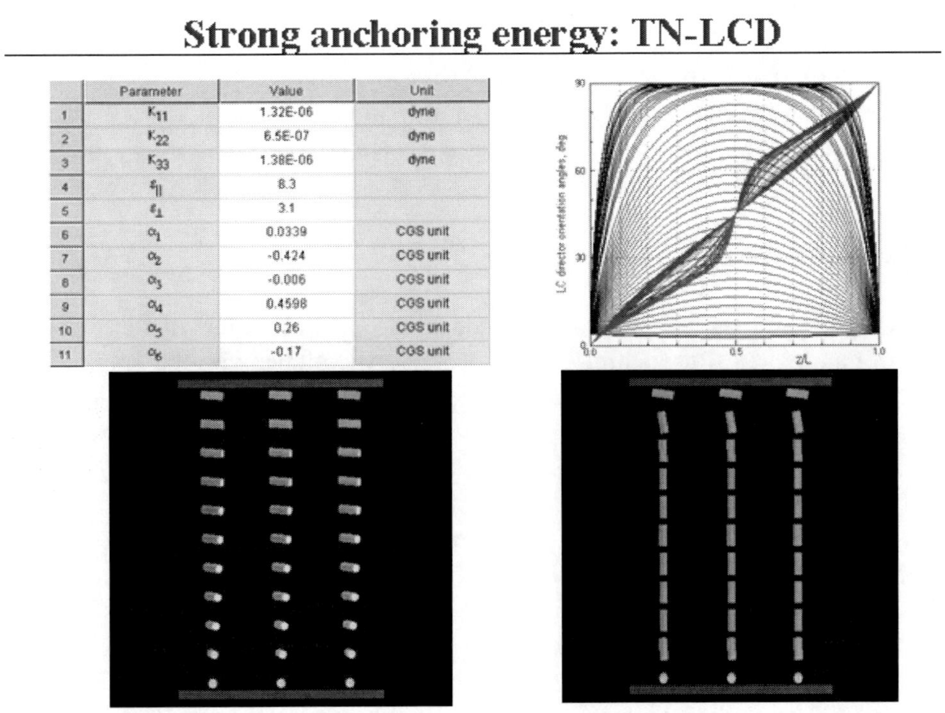

Figure 3.16. Evolution of the TN structure for a sufficiently high anchoring energy, 90^0 angle between LC directors keeps the same, even for a very high electric field.

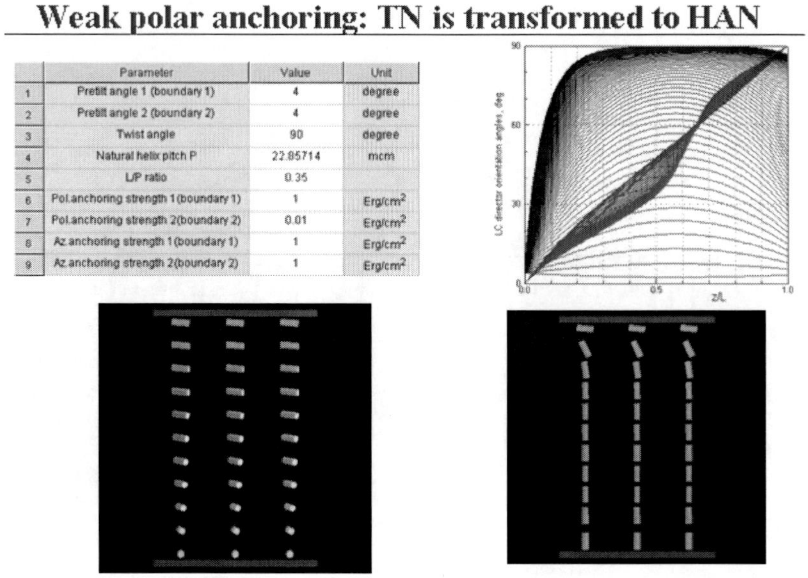

	Parameter	Value	Unit
1	Pretilt angle 1 (boundary 1)	4	degree
2	Pretilt angle 2 (boundary 2)	4	degree
3	Twist angle	90	degree
4	Natural helix pitch P	22.85714	mcm
5	L/P ratio	0.35	
6	Pol.anchoring strength 1(boundary 1)	1	Erg/cm^2
7	Pol.anchoring strength 2(boundary 2)	0.01	Erg/cm^2
8	Az.anchoring strength 1(boundary 1)	1	Erg/cm^2
9	Az.anchoring strength 2(boundary 2)	1	Erg/cm^2

Figure 3.17. TN structure, which still remain even for very high voltages in a strong anchoring case (Figure 3.16), can practically disappear for a low polar anchoring energy on one of the substrates (Figure 3.17). HAN structure occurs for very high voltages.

TN structure, which still remain even for very high voltages in a strong anchoring case (Figure 3.16), can practically disappear for a low polar anchoring energy on one of the substrates (Figure 3.17). Instead of TN another HAN structure occurs for very high voltages.

3.2.4. Dynamics of Twist Effect

The definition of the rise t_r (3.10) and decay t_{decay} (3.11) times of the twist effect are clarified in Figure 3.18. Responding to the square wave voltage with the amplitude V liquid crystal twist cell between parallel polarizers increases its brightness from the initial value B_0 to the value of the optical threshold of the twist effect $B_0+0.1(B_m-B_0)$ during the so-called "delay" time t_d. The rise time t_r corresponds to the further increase of the brightness up to the value of 90% from the difference between the initial and the maximum B_m value. After switching off the voltage the twist LCD brightness relaxes with a decay time t_{decay}. Qualitatively, the times for both the twist effect and the layer deformations are proportional to the viscosity and to the square of the thickness, and inversely proportional to an elastic constant of the liquid crystal material (3.11).Also the rise times (3.10) are inversely proportional to the difference $(V^2 - U_{tw}^2)$, where U_{tw} is a threshold voltage [1-3].

For the higher applied voltages, the edge of the twist effect oscillograms develops a characteristic bounce (there is no such bounce on the front edge) (Figure 3.19).

The effect takes place if the driving voltage is sufficiently high: $V>2U_{tw}$. This effect is accounted for theoretically by allowing for the backflow [1-3]. The point is that, in the initial stages of the relaxation, the maximum elastic torque is found in the region of a rapid change of the director deformation θ with the z-coordinate in the vicinity of the walls as in the ECB effect (Figure 3.20). The greatest rate of the reorientation of the director also occurs in this region, inducing the maximum associated flow of the liquid. This flow influences the director in the layer center in such a way that the director angle θ increases (to values above π/2). As relaxation proceeds, the backflow decreases, and the director can return to its initial state under the elastic forces. This corresponds to the maximum transmission of the cell (a bounce of the rear edge).

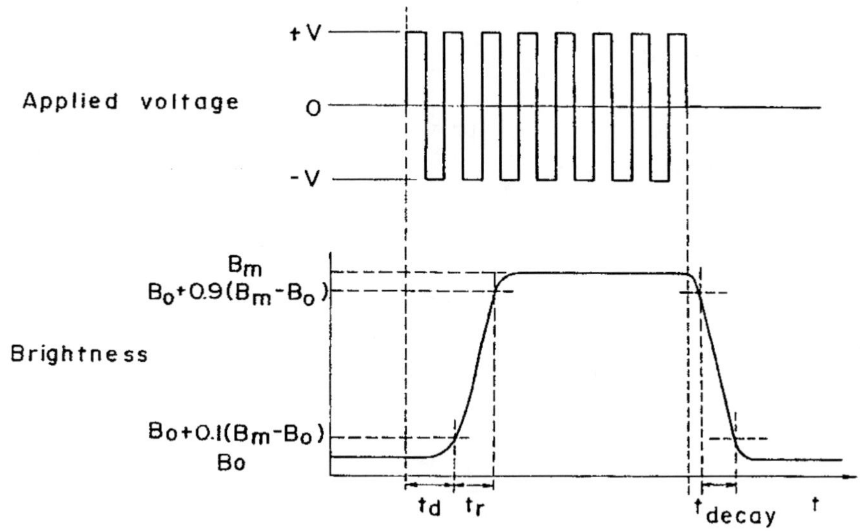

Figure 3.18. Typical oscillograms, which define the response time of the twist effect (a), The occurrence of the bounce on the decay of the TVC curve (b) [1-3].

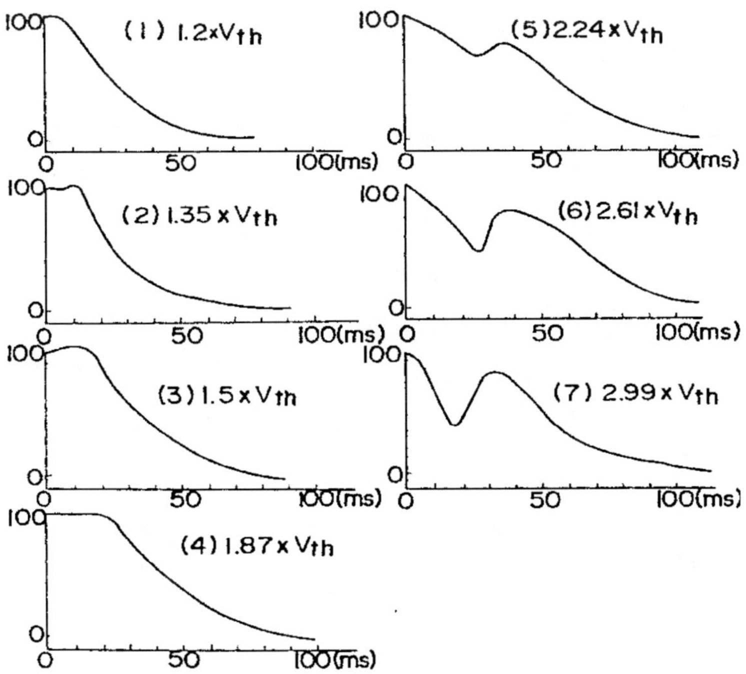

Figure 3.19. The occurrence of the bounce on the decay of the TVC curve [1-3].

There exists a possibility of decreasing the twist-effect relaxation times by applying a field with a frequency greater than that at which the dielectric anisotropy changes sign, as was discussed above for ECB effect. However, higher operating voltages and a limited operating temperature range prevent the dual frequency addressing scheme from finding an effective application.

3.2.5. New Developments

Let us analyze some new possible improvements of twist-effect LCDs, which recently makes it more attractive for electrooptical applications in Liquid Crystal Devices (LCDs).

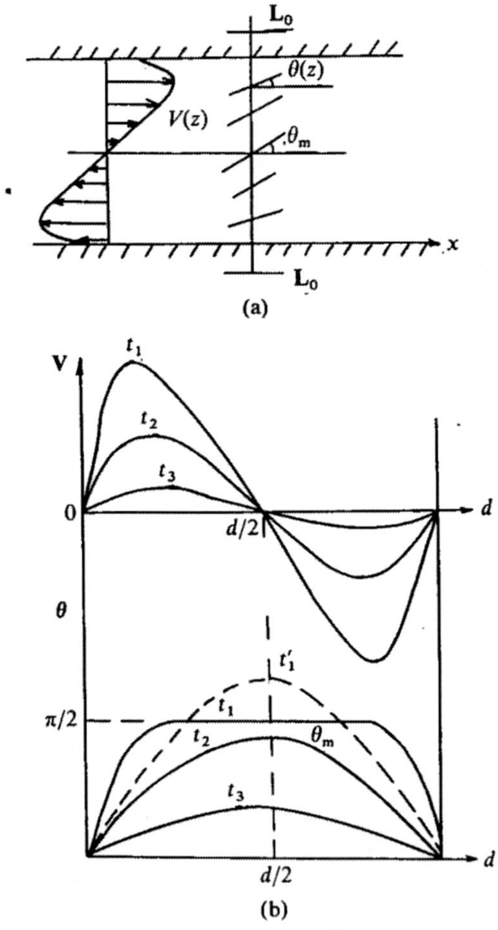

Figure 3.20. (a) Back-flow effect in the process of the director reorientation [1-3]. (b) The dependences of the velocity V and director angle θ on the coordinate z are given at different times: $t_1 < t_2 < t_3$. The dotted line shows possible non-monotonous relaxation of θ at high voltages: $t_1 < t'_1 < t_2 < t_3$ [1-3].

3.2.5.1. Effect of Weak Boundary Anchoring. Reverse Twist Mode

The effect of saturation, i.e., total reorientation of the twist director alignment to the homeotropic one at finite voltage values is possible. This effect could be used for the realization of memory states [1-3]. A considerable increase in the multiplexing capability of the twist cell at low anchoring energies is observed. The corresponding number of addressing lines is very sensitive to the form of the anchoring potential. For sufficiently small anchoring energies, the twist cell response reveals bistability and hysteresis [1-3]. It should be pointed out that the above mentioned is true only for cells with weak polar anchoring energies. Now the development of such systems is in progress. The azimuthal anchoring should be kept strong in all cases, otherwise the twist alignment can disappear at a certain critical thickness [1-3].

If the electrodes of a cell which have been pretreated by rubbing were further coated with a layer of surfactant (e.g., lecithin), the resulting orientation of the molecules of the nematic liquid crystal will be homeotropic and appears bright between parallel polarizers (Figure 3.6). In an electric field a nematic liquid crystal with a negative dielectric anisotropy must reorient itself into a twist structure, and transmission through the cell decreases to zero (reverse twist effect) [1-3]. This effect has been observed experimentally in comparatively thick cells ($d \approx 50$ microns). In cells with $d \approx 20$ microns, the final twisted state proves to be unstable and the nematic liquid crystal layer gradually transformed into a planar structure. The addition of small quantities of cholesteric liquid crystals to the initial nematic mixture stabilizes the final twisted structure and improves the electrooptical characteristics of the device. The electrooptical response of electrically induced twist nematic cells includes intensity oscillations observed both in the switching on and switching off regimes. These oscillations take place due to the variation of birefringence, which are not important in the usual twist effect.

Another reverse twist mode takes place, when the boundary conditions, that favors the twist configuration with a definite direction and the handedness of the chiral dopant in LC materials are opposite to each other [1-3]. Based on the LC molecular arrangement, low driving voltage and fast response times could be realized.

3.2.5.2. Improvement of Viewing Angles

One possibility of improving the viewing angles of the twist-effect is an application of the so-called “In Plane Switching” (IPS), when the electric field is applied parallel to the substrates plane. In this case it is possible to control the TVC of the twist cell mainly not by changing its phase retardation $\Delta\Phi$, but by effecting the angle φ_0 between the polarization plane of the incident light and the director axis (3.16). The dark state will correspond to the value of $\varphi_0=0$ and the angular dependence of the intensity will be absent as it is connected only with the phase retardation term $\Delta\Phi$ [1-3]. At the same time the coloration, i.e. the wavelength dependence of the transmission is also absent due to the same reason.

Another possibility of getting the wider viewing angles in twist effect is the so-called multi-domain alignment technique, i.e. dividing each twist cell to the two or four domains to observe the best side of the twist configuration, presented by each domain [1-3]. In some cases the multidomain configuration can be obtained by the non-fixed (amorphous) liquid crystal orientation on one of the substrates of the cell [1-3].

To improve the viewing angle dependence and contrast ratio of the twist cell, phase retardation plates are used [1-3]. Sometimes a second twist cell is placed after the first one with a 90^0 twist in the opposite sense. The two cells optically compensate for each other when place so that their directors are perpendicular on the facing surfaces. Thus the double-layered twisted device appears black between crossed polarizers for all wavelengths and has better viewing characteristics at oblique incidence.

3.2.5.3. Stress Splay Twist Mode

Special "stress" configuration of TN-LCD can both decrease the driving voltage (power consumption) and improve response time (Figure 3.21) [10, 11]. The twist sense of TN-LCD is determined by the directions of pretilt angles on both alignment layers. In commmmercially available TN-LCDs, the chiral material is used to stabilize the LC twist sense. The LC materials for TN-LCDs possess the twisting property of the same direction as the one determined by the combination of pretilt angle directions. On the contrary, by adding the chiral reagent of which twist direction is opposite to the one determined by the combination of pretilt angle directions, the splayed twist state is formed (Figure 3.21). The stability of "stress" TN state structure depends largely on the pitch length and pretilt angle. The larger pitch length and pretilt angle can stabilize the new mode.

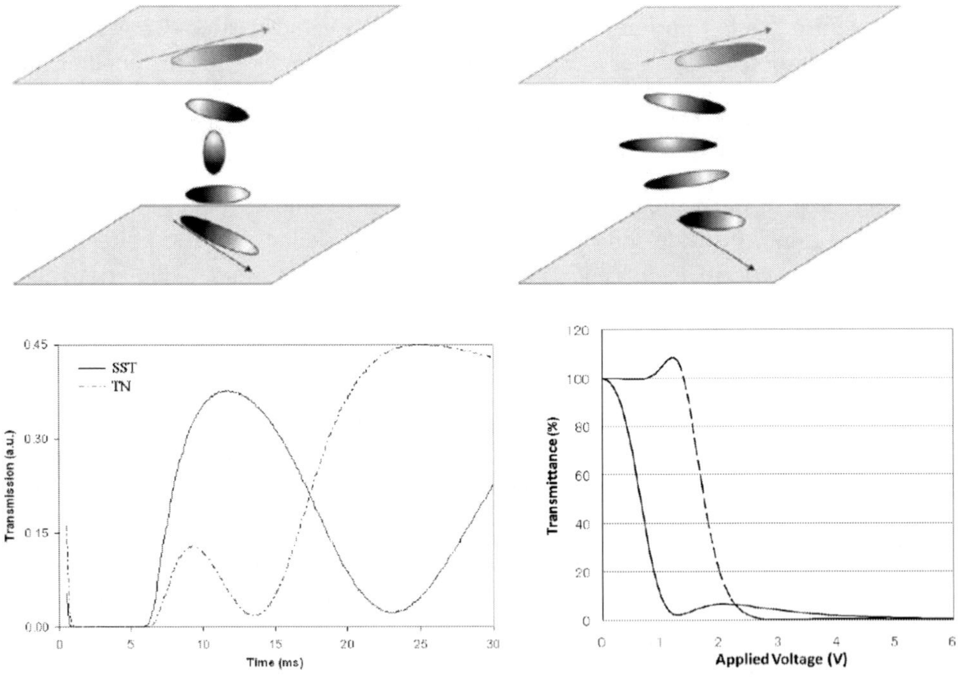

Figure 3.21. Stress Splay Twist (SST) [11] or Reverse TN (RTN) [10] Mode. Top: common TN (left), SST/RTN mode (right). Bottom: Comparison of response time (left) [11] and applied voltage (right) [10] between common TN and SST/RTN mode.

3.3. Supertwist Effect

The picture of supertwist LC configuration [1-3] is shown on Figure 3.22. A small voltage difference transforms a highly twisted (supertwist) LC state to almost homeotropic configuration.

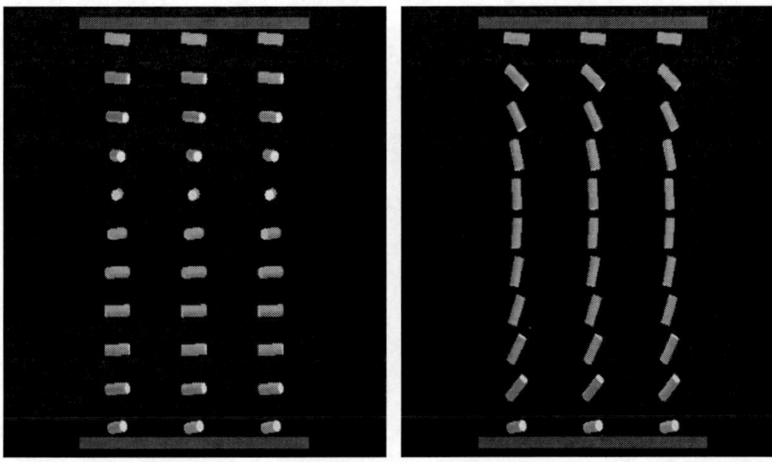

Figure 3.22. Director deformation in supertwist LC cell. Left: "off" state, right: "on" state. The voltage difference between "off" and "on" states is rather small in comparison with TN LC cell.

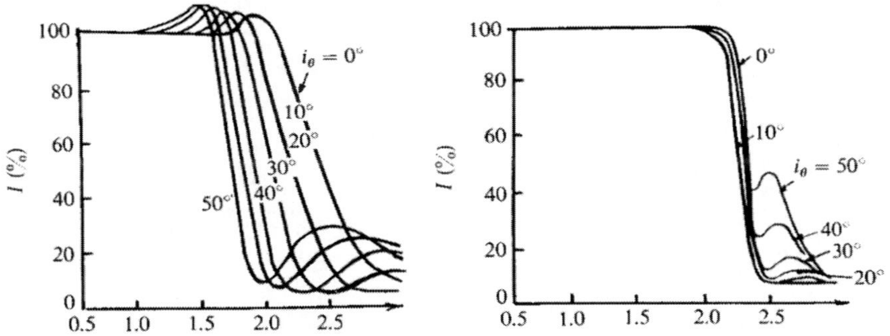

Figure 3.23. Transmission-voltage characteristics: (a) -90^0 twist structure; (b)- a 200^0 supertwisted cell. The light incidence angles i_θ are shown on the curves, $i_\varphi=180^0$ [1-3].

The transmission –voltage characteristics in supertwist LC cell becomes steeper, while the angle dependence is getting smoother, than for TN LC cell (Figure 3.23).

3.3.1. Various Supertwist Modes

The general scheme of realization of supertwist LC cells is shown in Figure 3.24 [1-3]. Here the general scheme of supertwist display geometry shows the input and output orientations of the molecules or their projections (L_{in}, L_{out}) as well as the input and ouput orientation of the polarizers (P_{in}, P_{out}). The angle φ_m is the supertwist angle, β and η define the location of the polarizers: η is the angle between polarizers, and β-angle of the first polarizer with respect to the director on the front substrate. As seen from Figure 3.24 various supertwist geometries could be obtained by altering the supertwist angle φ_m and polarizer angles β, η. Besides that we can change such parameters of the supertwist mixtures as optical path difference Δnd and the director pretilt at the boundaries θ_0. Various realizations of supertwist structures for passively addressed high information content LCDs are shown in Table 3.1[1-3]. New methods for the realization of the electrooptical effects in supertwist LC cells such as supertwist nematic (STN) or optical mode interference (OMI) avoid certain limitations and disadvantages which observed in supertwist birefringence (SBE) LC cells. For example, the boundary tilt angles for preventing the appearance of domain structures are not as large as in SBE case and the requirements to thickness nonuniformity become softer. Consequently, the manufacturing of LC displays becomes easier. STN and OMI mixtures have the same or even better steepness as the SBE prototype, which means addressing of the larger number of lines in LC display with a passive addressing [1-3].

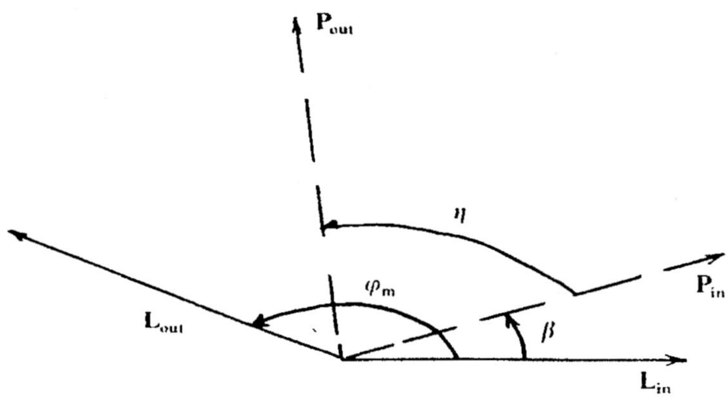

Figure 3.24. The general scheme of supertwist display geometry shows the input and output orientations of the molecules or their projections (L_{in}, L_{out}) as well as the input and ouput orientation of the polarizers (P_{in}, P_{out}). The angle φ_m is the supertwist angle, β and η define the location of the polarizers [1-3].

The Optical Mode Inteference (OMI) effect provides a weak-wavelength dependence of the transmission in the visible region [13]. The effect requires a low optical path difference, which leads to a strong interference of two polarization modes when propagating through an OMI cell. (Mauguin's waveguide regime (3.22) is not valid). The transmission spectra of the OMI cell enable us to realize the black and white appearance of the two display states, which is impossible in STN or SBE cells. The electrooptical characteristics of the OMI-cells are much more tolerant to the cell gap nonuniformity, than in the STN case and are less temperature dependent. However, one of the main disadvantages of the OMI display is low brightness in the off-state. For instance, if we take the brightness of two parallel polarizers equal to 100%, then the "off" states of the 90^0 twist cell and the 240^0 STN cell would correspond to 95% and 64%, respectively, while the brightness of the 180^0 OMI cell does not exceed 40% [14]. However, the brightness of OMI displays can be greatly improved up to 77% by increasing the twist angles up to 270^0, if we make an appropriate choice of Δnd and the angular position of polarizers [15]. Besides that OMI displays possess the response times about 1.5 times lower, than STN-LCDs for the same value of the contrast and multiplexing capability. Surprisingly the viewing angles of OMI displays for high-information content screens are not much wider, than in STN case, despite the lower values of Δnd. Thus OMI displays are competitive with STN displays for high information content passively addressed LCDs.

Table 3.1. Supertwist structures for passively addressed high information content LCDs [1-3]

Supertwist mode	Supertwist angle	Thickness-to-pitch ratio, d/P	Optical path length, Δnd, μm	Polarizer angles β, η	Pretilt angles on the substrates, θ_0
Supertwist birefrigent effect (SBE) [12]	270^0	0.75	0.85	-32.5^0, 65^0	28^0
Supertwist nematic effect (STN)	240^0	0.66	0.82	45^0, 60^0	4^0 -6^0
Optical mode interference (OMI) effect[13]	180^0	0.3	0.46	0^0, 90^0	2^0 -3^0
Optical mode interference (OMI) effect	240^0	0.43	0.54		4^0 -6^0

3.3.2. Supertwisted LCDs with Improved Characteristics

3.3.2.1. Parameter Space Approach

Parameter Space Approach was proposed in [16] to optimize the properties of LCD with various twist angles, birefringence and location of the polarizer and analyzer for transmissive and reflective LCD. The method can be successfully applied for the optimization of STN-LCD (Figure 3.25). If we choose various angles of the polarizer α, then the positions of the maximum transmissions, i.e. the most suitable values of dΔn and φ will be changed accordingly (Figure 3.26) [16,17].

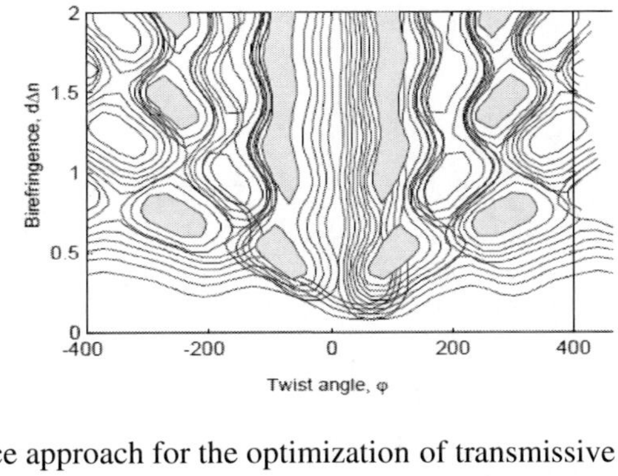

Figure 3.25. Parameter space approach for the optimization of transmissive and reflective LCD. Four independent parameters are used: polarizers angles *α, γ*, phase retardation dΔn, twist angle φ. LCD transmission T is shown as a function of a phase retardation dΔn and twist angle φ. The optimal values of phase retardation dΔn and twist angle φ corresponds to the position of the maximum transmission [16].

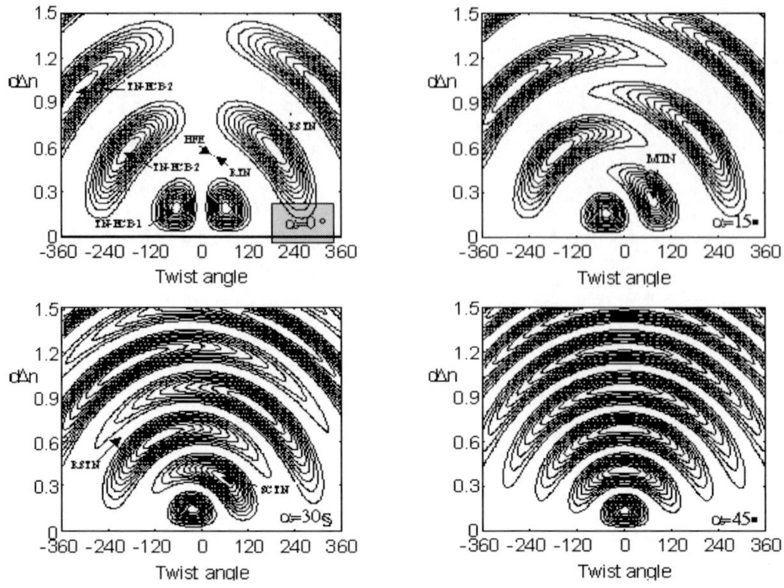

Figure 3.26. The positions of maximal transmission in a parameter space approach as functions of the polarizer angle α, phase retardation dΔn, and twist angle φ. LCD transmission T is shown as a function of a phase retardation dΔn and twist angle φ. The optimal values of phase retardation dΔn and twist angle φ correspond to the position of the maximum transmission [16,17].

Let us also note, that the optimal values of dΔn and φ were calculated for only one value of the light wavelength, which corresponds to the maximum sensitivity of the human eye (λ=550nm, green light). This does not mean that these values are optimal for the whole spectrum of visible light. We will consider this more in detail below. As seen from Figure 3.26, various optimal phase retardations dΔn and twist angles φ corresponds to different modes: like TN-ECB (twist nematic - electrically controlled birefringence), MTN (mixed twist nematic), RSTN (reflective supertwist nematic), SCTN (self compensated twist nematic), HFE (hybrid field effect) etc. Table 3.2 shows the mode name as well as the optimal values of α, dΔn and φ for the observation of these mode in normally black (NB) and normally white (NW) states.

Table 3.2. Various optimal modes for LCD operation [16,17]

Acronym	Name	Twist angle	Polarizer angle	Retardation	
				NW	NB
HFE	Hybrid field effect	45°	0°	0.362	0.968
RTN	Reflective twisted nematic	54°	0°	0.359	0.9515
TN-ECB	TN - electrically controlled birefringence	63.7°	0°	0.354	0.935
MTN	Mixed mode twisted nematic	~80°	~20°	0.45	0.864
SCTN	Self compensated twisted nematic	60°	30°	0.61	0.943

3.3.2.2. Film Compensated STN (FSTN)

One of the main goals is to attain a black-white switching of a supertwist cell with a high contrast ratio and an acceptable brightness. At the same time, we need to keep a sufficiently high steepness of the transmission –voltage curve (TVC) to enable a sufficient resolution of STN LCD. The possible solution of the problem is to use phase retardation plates in combination with an STN display. The idea is to compensate phase difference induced by a supertwisted structure. The supertwisted displays with one and two phase retardation plates were reported [1-3]. The orientation of the retardation plates and their optical path differences are optimized to provide both black and white switching and wide viewing angles. The double plates provide better achromatic appearance and contrast ratio than a single plate, especially when they are placed on both sides of the STN cell. Thin polymer films (polycarbonate, polyvinyl alcohol, etc.) are used now, as phase retardation plates. Biaxial compensator films and optically negative polymeric films composed of discotic molecules were also developed [18]. The contrast ratios of the STN-LCD with negative birefringence films achieve 100:1 for the normally incident light. However, it is very difficult to realize such a film with a uniform phase retardation over a large surface area.

Typical interface of a high resolution FSTN-LCD with phase retardation plates, and antireflective layers is shown in Figure 3.27. The calculations were made using MOUSE-LCD software [4]. The multiplexing duty ratio N=32:1 (number of addressing lines can be 2N=64 [1-3]), contrast ratio is more than 300:1 for a normally incident light.

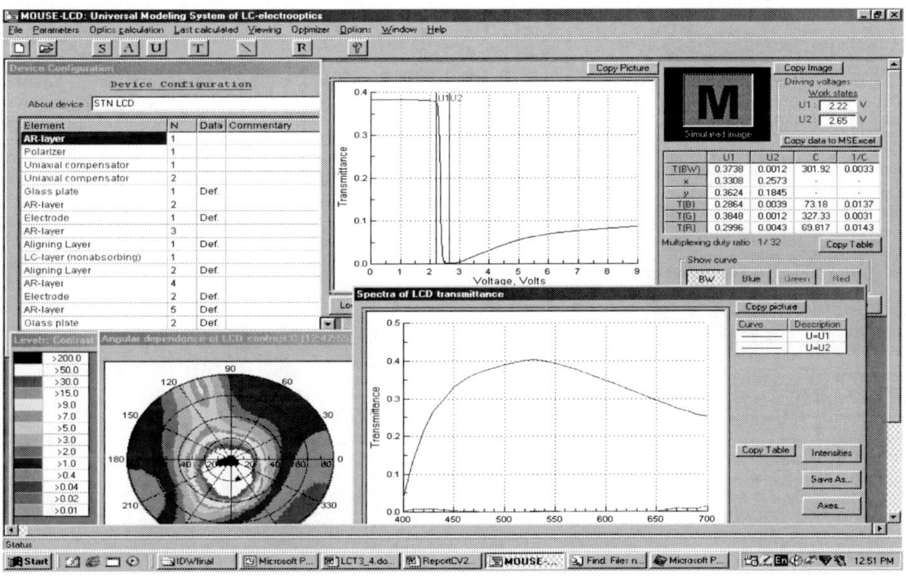

Figure 3.27. FSTN LCD, optimized by two subsequent operations: (i) parameter space; (ii) insertion of phase retarders and anti-reflective layers.

3.4. Electrooptical Modes in Cholesterics

3.4.1. Selective Reflection Band

Various cholesteric textures (planar, focal –conic, homeotropic) are shown in Figure 2.23. Further we will consider the electrooptic switching between these textures more carefully. Now let us summarize the basic optical properties of a cholesteric planar texture for the light normally incident to this texture (Figure 2.23) [1-3].

1. There exists one resonance band (selective reflection band), i.e. a full reflection of light with the circular polarization having the same handedness as the pitch of the helix and a corresponding total transmission of the light with an opposite circular polarization (Figure 2.23,a). The band has rather flat top with a center at

$$\lambda_0=R_0\langle n\rangle, \tag{3.28}$$

where $\langle n\rangle=\sqrt{(n_\|^2+n_\perp^2)/2}$. The higher orders of reflection are forbidden.

2. The spectral width of this reflection $\Delta\lambda$ is proportional to the optical anisotropy $\Delta n= n_\| - n_\perp$ of a cholesteric liquid crystal

$$\Delta\lambda=R_0\,\Delta n \tag{3.29}$$

3. On each side of the selective reflection band there are regions of strong rotation of a plane polarized light.

Schematic representation of wavelength dependence of optical rotatory power and circular dichroism in cholesteric systems are shown in Figure 3.28, b and Figure 3.28,c respectively.

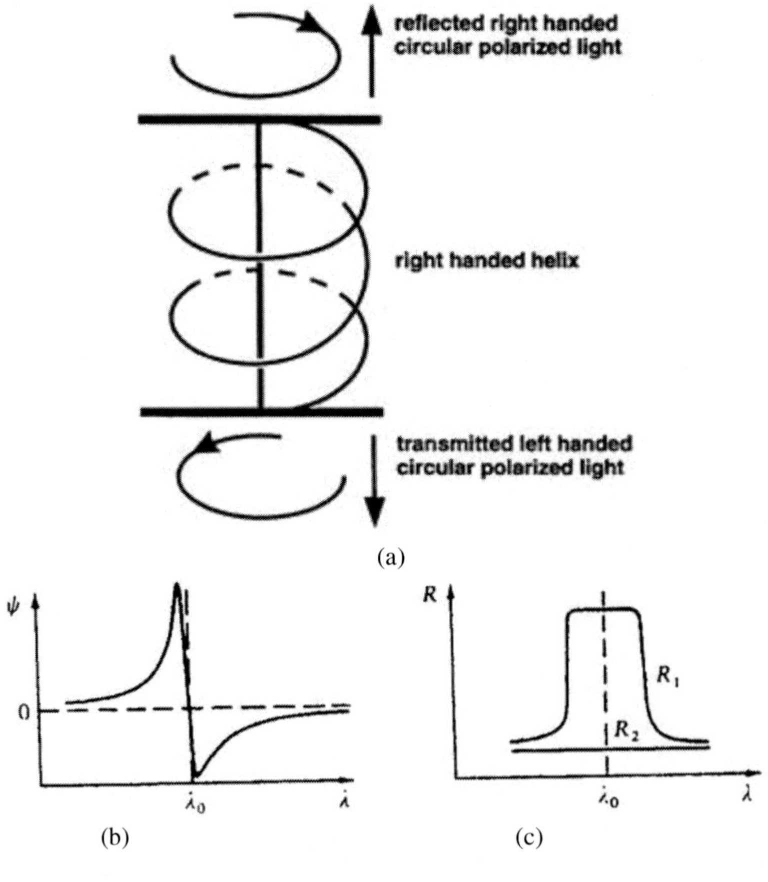

Figure 3.28. Optical properties of a cholesteric planar texture. (a)-selective reflection of a circularly polarized light; (b)- optical rotatory power; (c)-circular dichroism [1-3].

3.4.2. Unwinding of a Cholesteric Helix

Due to the nonzero dielectric anisotropy $\Delta\varepsilon \neq 0$ (2.9) and flexoelectric coupling the cholesteric LC interacts with the external electric field resulting in the two main processes: unwinding of the cholesteric helix and linear flexoelectric switching .

The transformation (unwinding) of the cholesteric helix in a magnetic field is shown in Figure 3.29 [1,2]. The process consists of (i) increasing helix pitch in intermediate field values $0<H<H_{th}$; (ii) divergence of helix pitch at the threshold vavue $H=H_{th}$. The corresponding magnitude of the electric field [1-3]:

$$E_{th}=\pi^2/R_0(4\pi K_{22}/\,\Delta\varepsilon)^{1/2} \tag{3.30}$$

The dynamics of helix unwinding in cholesterics is described by the characteristic rise time τ_{rise} and decay or relaxation time τ_{rel} according to the following relations:

$$\tau_{rise}= \gamma_1/(\Delta\varepsilon E^2/8\pi- K_{22}q^2),\ \tau_{rel}= \gamma_1/\,K_{22}q^2 \tag{3.31}$$

As follows from (3.31) the relaxation times of the helix diverge near the threshold field:

$$q=2\pi/R \Rightarrow 0 \text{ for } R \Rightarrow \infty \text{ at } E=E_{th}.$$

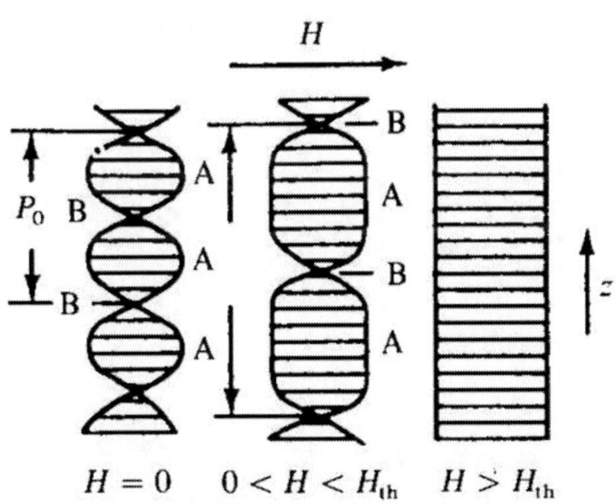

Figure 3.29. Unwinding of the cholesteric helix in a magnetic field perpendicular to the helix axis [1,2].

3.4.3. Linear Flexoelectric Effect

We should also allow for flexoelectric polarization (Figure 3.30)

$$\mathbf{P}_f= e_{11}\mathbf{L}\ \mathrm{div}\ \mathbf{L} - e_{33}(\mathbf{L} \times \mathrm{curl}\mathbf{L}) \tag{3.32}$$

where e_{11} and e_{33} are flexoelectric moduli, which gives rise to the following term g_f in the density of the nematic free energy:

$$g_f = -(\mathbf{P}_f\mathbf{E}) \tag{3.33}$$

The flexoelectric term is connected with the interaction of the LC director deformation and is complementary to the usual elastic term, which does not depend on the electric field (2.15). If a cholesteric axis is parallel to the plane of the cell the flexoelectric coupling leads to the reversible distortion of the structure with the characteristic time in the order of hundreds of microseconds [1-3]. For the values of the electric field sufficiently smaller, than the threshold $E<<E_{th}$ the deformation amplitude of the helix axis is:

$$\varphi(E) = \arctan(e_f E R_0/2\pi K) \tag{3.34}$$

i.e. changes its sign simultaneously with the field E. (Here e_f and K are average flexoelectric and elastic coefficients respectively, R_0 -helix pitch.) It results in the corresponding modulations of the intensity of the light on the output of LC cell placed between crossed polarizers. The problem is not to disturb the helix pitch too much as the electrooptic response may deviate from the linear character due to the helix unwinding effect. We will not consider here the appearance of dielectric, flexoelectric and other modulated scattering structures in cholesterics, which are clearly described elsewhere [1-3]. Such structures are parasitic and must be avoided by a special choice of LC parameters, cell construction (e.g. increased tilt angle) and driving regime.

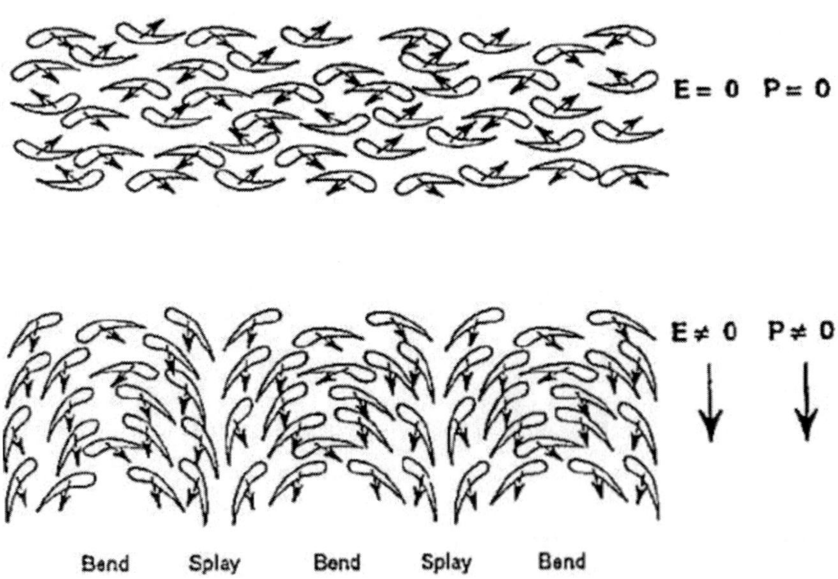

Figure 3.30. Flexoelectric deformation in a planar cholesteric structure.

3.4.4. Reflective Cholesteric Structures

The reflective cholesteric textures were intensively investigated [19]. The basic structures were so-called Polymer-Stabilized Cholesteric Textures (PSCT) with a stabilization by dispersing a low concentration of polymer inside the cholesteric liquid crystal. The electric field can switch between the three main PSCT configurations [1-3] (Figure 3.31).

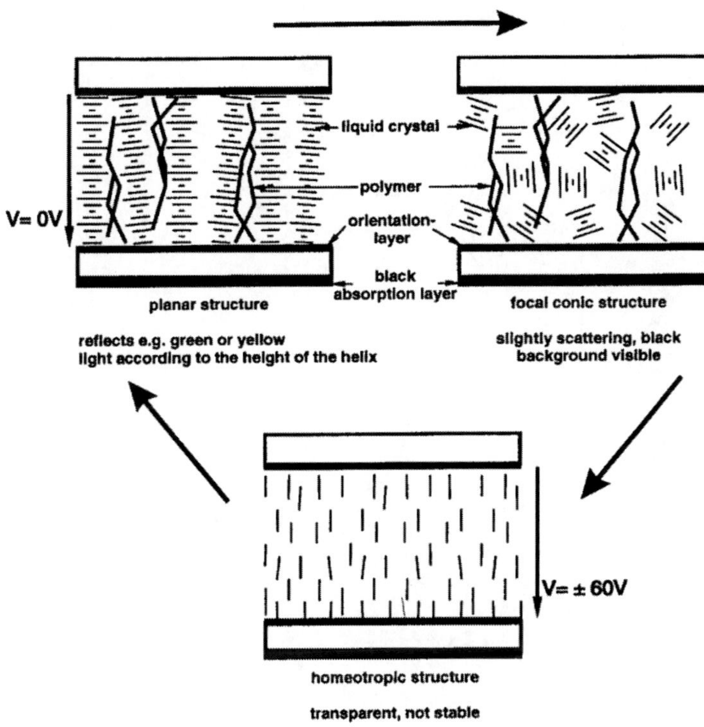

Figure 3.31. Switching between three cholesteric textures in PSCT configuration [1-3].

The planar cholesteric structure is prepared by UV-curing of the mixture of a cholesteric liquid crystal, a monomer, and a photoinitiator with a simultaneous application of electric field to achieve the homeotropic orientation of LC molecules together with a monomer for forming a proper polymer network [19]. The competition between the polymer network, which favors the homeotropic LC alignment and the surfaces treated for a homogeneous LC orientation results in a multi-domain PSCT structure (Figure 3.31). The cell is then coated with a black absorbing layer to enhance the contrast between the two PSCT states.

The planar PSCT structure exhibits Bragg reflection of green or yellow light in accordance with the formula (3.28). The texture is stable in the absence of the applied field, so no power is needed to keep the image, when it is written (intrinsic memory effect). The image is readable in both indoor room lighting and outdoor sunlight conditions. The contrast is very good for wide viewing angles and no polarizers are needed. The wavelength of the reflected light is easily controlled by developing chiral nematic mixtures with the temperature stabilized pitch R, polymer network and/or surface alignment layers.

The PSCT planar structure with a positive dielectric anisotropy $\Delta\varepsilon > 0$ can be electrically driven to a focal-conic one by applying a short pulse of a sufficiently high voltage (about

±40v for a 7 microns thick sample). In the focal-conic structure cholesterics become slightly scattering making visible black background, specially prepared for the purpose (Figure 3.31). The contrast and viewing angles of the PSCT cells are defined by a Bragg color reflection in the planar state and slightly scattered state near the dark background (Figure 3.32)[1-3]. The viewing angle in the focal conic state is large partially due to the random orientation of the helixes and different size of domains (Figure 3.31). The superiority of the PSCT LCD viewing angle characteristics over reflective STN-LCDs was clearly shown [19].

The new developments in reflective cholesteric displays include photosensitive cholesteric displays [20]: when irradiated, the resulting conformation variation of the dye in the LC bulk affects the helical twisting power and changes the pitch of the cholesteric, therefore, changing the peak reflected wavelength of the display.

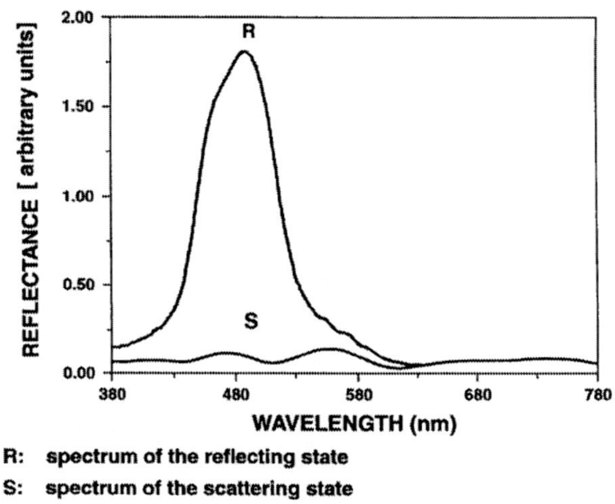

Figure 3.32. Spectral and viewing angle characteristics of PSCT LCDs. Spectral characteristics in the two different states [1-3].

3.5. ELECTROOPTIC EFFECTS IN FERROELECTRIC LC

3.5.1. Basic Physical Properties

3.5.1.1. Symmetry

The symmetry of the ferroelectric smectic C* phase corresponds to the polar symmetry group C_2, Figure 3.33, so that when going along the z-coordinate parallel to a helix axis and perpendicular to the smectic layers the director **L** and the polarization vector **P**, directed along the C_2 axis, rotate such as

$$\mathbf{L}(z + R) = \mathbf{L}(z), \mathbf{P}(z + R) = \mathbf{P}(z), \tag{3.35}$$

i.e., the helix pitch R is equal to a spatial period of the FLC structure. In the absence of external fields the FLC equilibrium helix pitch is Ro and the average polarization of the FLC volume is equal to zero, Figure 3.33.

3.5.1.2. Main Physical Parameters

The main physical parameters which define FLC electrooptical behavior are:

(i) tilt angle, θ;
(ii) spontaneous polarization, P_s;
(iii) helix pitch, R_0;
(iv) rotational viscosity, γ_φ;
(v) dielectric anisotropy, $\Delta\varepsilon$;
(vi) optical anisotropy, Δn;
(vii) elastic moduli; and
(viii) anchoring energy of the director with a solid substrate.

Let us briefly characterize each of these parameters.

Tilt Angle

The value of the tilt angle could vary from several degrees to $\theta \approx 45^0$ in some FLCs. Usually, in electrooptic FLC materials, $\theta \approx 22.5^0$ in an operating temperature range. However, for some electrooptical applications it is desirable to have the value of θ as high as possible. The temperature T dependence near the phase transition point T_c to the more symmetric LC phase (Smectic A, Nematic or Chiral Nematic)

$$\theta \approx (Tc - T)^{1/2} \tag{3.36}$$

is typical of second-order phase transitions.

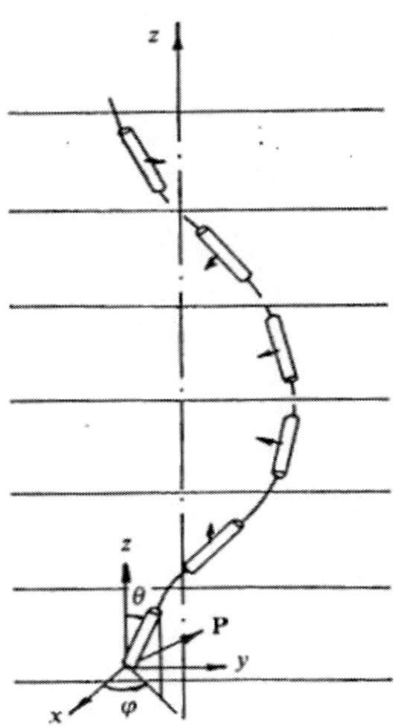

Figure 3.33. Ferroelectric liquid crystal (FLC) structure .

Spontaneous Polarization

The value of the spontaneous polarization depends on the molecular characteristics of an FLC substance itself and the achiral dopant introduced into the matrix, and can vary from 1 to 200 nC/cm^2 or even more [1-3]. The value of the spontaneous polarization is one of the main FLC characteristics, which define FLC electrooptic response.

Rotational Viscosities

The switching times in the electrooptical effects in FLCs is defined by the rotational viscosity γ_φ which characterizes the energy dissipation in the director reorientation process. According to the FLC symmetry, two viscosity coefficients should be taken into account, γ_θ and γ_φ, which determine the corresponding response rates with respect to the director angle φ (Figure 3.33). The relevant dynamic equations take the form [1-3]:

$$\gamma_\varphi\, d\varphi/dt - P_S E \sin\varphi = 0,\ \tau_\varphi = \gamma_\varphi / P_S E, \tag{3.37}$$

where τ_φ -characteristic response times for FLC director. The viscosity coefficient γ_φ can be rewritten as

$$\gamma_\varphi = \gamma_\varphi' \sin^2\theta \tag{3.38}$$

where γ_φ' is independent of the angle θ. According to (3.38), $\gamma_\varphi \Rightarrow 0$ for $\theta \Rightarrow 0$, i.e. γ_φ is very low for small tilt angles θ.

Near T_c one can change the θ angle, e.g. by applying electric field E, this effect is known as electroclinic effect [21]. The rotational viscosity γ_φ could be estimated from the experimental dependence of the electrooptical response as follows [22]:

$$\gamma_\varphi = P_S E \tau_\varphi$$
$$\tau_\varphi = (t_{90} - t_{50})/\ln\sqrt{5} \tag{3.39}$$

where t_{90} and t_{50} are the corresponding times for 90% and 50% transmission from the maximum level (Figure 3.34).

One of the possible ways to increase the switching rate in FLCs includes minimizing the viscosity with a simultaneous rise in polarization. However, this way is not very promising, because: (i) the increase in polarization is hardly compatible with the lower values of rotational viscosity [22]; and (ii) the repolarization component of the a.c. current grows rapidly with the value of polarization, i.e.

$$i_P \approx P_S/\tau_\varphi \approx P_S^{\ 2}/\gamma_\varphi \tag{3.40}$$

which is always undesirable for applications.

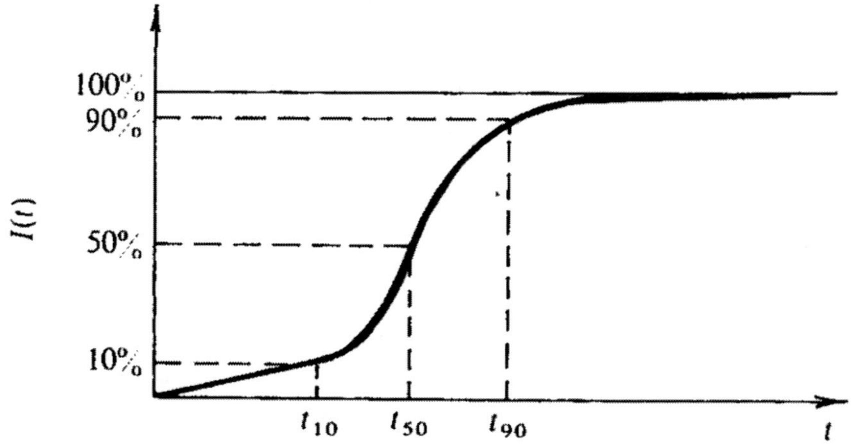

Figure 3.34. Dynamics of FLC cell transmission [1-3].

Helix Pitch

The helix pitch value R_o is easily controlled by varying the concentration of a chiral dopant in a smectic C matrix. To provide variation of the helix pitch from 0.1 μm to 100 μm and more, we should have a chiral dopant with a high twisting power and good solubility. At the same time, the chiral dopant should not depress the smectic C temperature range [1-3]. Moreover, the growth of polarization should be more pronounced than that of the rotational viscosity γ_φ with an increasing concentration of the chiral dopant. The inverse pitch of the FLC helix, for reasonably large concentrations of the helix of the chiral dopant C_D, is proportional to the concentration:

$$R_0^{-1} \propto c_D \tag{3.41}$$

Dielectric and Optical Properties

It is of importance for applications to have negative dielectric anisotropy $\Delta\varepsilon = \varepsilon_{\|} - \varepsilon_{\perp} < 0$, which stabilizes the relevant director structures [1-3]. Usual values of the negative dielectric anisotropy $\Delta\varepsilon$ in FLC mixtures are between –0.5 and –2 in kHz region decreasing at the higher frequencies [1-3].

The effect of non-zero dielectric anisotropy can be taken into account, inserting the additional term into the equation (3.37) of the azimuthal director motion in the electric field E[1-3]:

$$\gamma_\varphi \, d\varphi/dt = P_S E \sin\varphi + (\Delta\varepsilon/4\pi) E^2 \sin^2\theta \sin\varphi \cos\varphi \tag{3.42}$$

The dielectric tensor of an FLC at optical frequencies could be regarded as uniaxial [1-3]. Then the FLC possesses only two refractive indices, $n_{\|}$ along the director and $n_{\perp}$ perpendicular to it. The electrooptical behavior of an FLC is mainly defined by the optical anisotropy $\Delta n = n_{\|} - n_{\perp}$. The birefringence value Δn can be obtained from electrooptical measurements, where the dependence of the transmitted intensity on the phase factor $\Delta n d/\lambda$ (d is the cell thickness and λ is the light wavelength) is used [1-3]. The dispersion law for the optical birefringence follows the well-known Cauchy formula:

$$n(\lambda) = \Delta n(\infty) + C/\lambda^2, \ C=\text{const}. \tag{3.43}$$

Elastic Properties and Anchoring Energy

The elastic properties of FLCs are usually discussed using the density of the elastic energy g_{el} as follows [1-3]:

$$g_{el} = ½[K_{11}(\text{div}\mathbf{L})^2 + K_{22}(\mathbf{L}\ \text{curl}\ \mathbf{L} - t)^2 + K_{33}(\mathbf{L}\text{xcurl}\ \mathbf{L} - \mathbf{b})^2], \tag{3.44}$$

where t and **b** is spontaneous twist and bend of the FLC director **L,** K_{ii} are FLC elastic moduli. The value of t>0 or t<0 for the right and left handed FLCs respectively. The general continuum theory of FLC deformations must take into account not only deformations of the FLC director **L**, but also possible distortions of the smectic layers, which can be described in terms of the variation of the layer normal **ν** [1-3].

The FLC free energy should also include the surface terms, which are polar $w_p = -W_p(\mathbf{P}\boldsymbol{\nu})$ and dispersion $w_d = -W_d(\mathbf{P}\boldsymbol{\nu})^2$ contributions, where W_p and W_d are the corresponding anchoring strength coefficients, **v** is the layer normal, and **P** is the FLC polarization [23]. Thus the total free energy F_d of the FLC director deformations

$$F_d = \int_v g_{el}\, d\tau + \int_s (w_p + w_d) d \tag{3.45}$$

In the presence of the external field, the total free energy includes the energy of the director deformations F_d and the energy F_E of the interaction of the ferroelectric phase with the field **E**:

$$F = F_d + F_E = F_d + \int_v [-(\mathbf{PE}) - \mathbf{DE}/8\pi]\, d\tau. \tag{3.46}$$

i.e. consists of a sum of the volume (v) and surface (s) terms. Various textures of FLC are observed [1-3]: helical state in sufficiently large thickness $d>>R_0$, where R_0 is a helix pitch, d-FLC cell thickness; uniform states with $R_0>d$ due to the parallel orientation order effect on the substrates ("up" and "down" states differ by the direction of the FLC polarization **P**) and twisted state for large values of $W_p/W_d >>1$, which favors anti-parallel surface aligning of the FLC director.

Anchoring energy of FLC can be determined either (i) by measuring the width of the coercivity loop ΔV in the P(E) dependence in a static field according to the relation $\Delta V= 8W_d/P_s$ or (ii) by measuring the free relaxation times τ_r of the FLC director to the bistable states $\tau_r = \gamma_\varphi d/ 4W_d$, where γ_φ is FLC rotational viscosity [24]. The results of these two methods coincide with each other with the accuracy of 30%.

3.5.2. Electrooptic Effects in FLC Cells

3.5.2.1. Clark-Lagerwall Effect

Let us consider the main electrooptical phenomena in FLC. The best known is the Clark-Lagerwall effect [25], which results in director reorientation from one bistable state to the other when an external electric field changes its sign (Figure 3.35). In this case, the FLC layers are perpendicular to the substrates and the director moves along the surface of a cone whose axis is normal to the layers and parallel to the cell substrates. In each final position of its deviation the director remains parallel to the substrates, thus transforming the FLC cell into a uniaxial phase plate. The origin of electrooptical switching in the FLC cell is the interaction of the polarization **P** perpendicular to the director with the electric field **E**. The maximum variation of the transmitted intensity is achieved when the FLC cell is placed between crossed polarizers, so that the axis of the input polarizer coincides with one of the final director positions. The total angle of switching equals the double tilt angle θ (Figure 3.35). The Clark-Lagerwall effect is observed in the so-called Surface Stabilized FLC structures (SSFLC). In SSFLC cells d < Ro and we come to the situation where the existence of the helix is unfavorable, i.e., the helix is unwound by the walls.

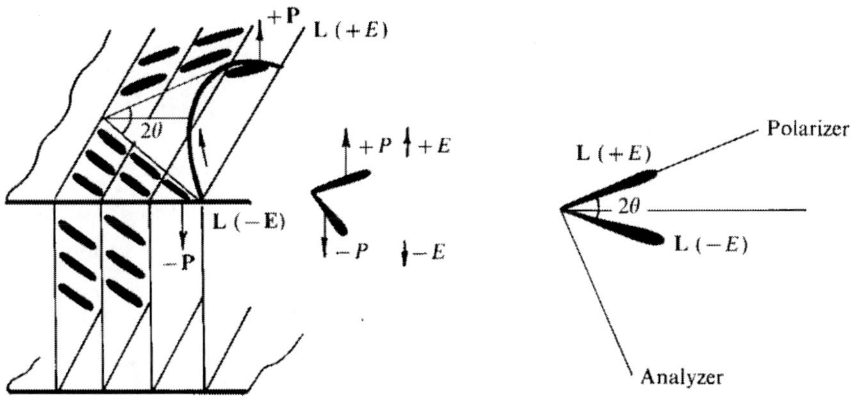

Figure 3.35. Clark-Lagerwall effect in FLC cell.

The variation of the azimuthal director angle φ in the Clark-Lagerwall effect (Figure 3.35) is described by the equation of the torque equilibrium (3.42), which comes from the condition of the minimum of the FLC free energy (3.46). The total model of FLC reorientation in a "book-shelf" geometry (Figure 3.35) with appropriate boundary conditions is:

$$\gamma_\varphi\, d\varphi/dt = P_S E \sin\varphi + (\Delta\varepsilon/4\pi) E^2 \sin^2\theta \sin\varphi \cos\varphi,$$
$$K\partial\varphi/\partial x + W_p \sin\varphi \pm W_d \sin 2\varphi \big|_{z=o,d} = 0 \qquad (3.47)$$

For typical values of polarizations $P_s \approx 20$ nC/cm^2, the driving fields $E \approx 10$ V/μm, and the dielectric anisotropy $\Delta\varepsilon \approx 1$, we have

$$|\Delta\varepsilon E/4\pi| < P_s, \qquad (3.48)$$

and, consequently, the dielectric term in (3.47) could be omitted. If the inequality (3.48) is invalid, what occurs for sufficiently high fields, then for $|\Delta\varepsilon E/4\pi| \propto P_s$ the response times of the Clark-Lagerwall effect sharply increase for positive Δε values [1-3]. Experiment shows that for negative values of Δε the slope of the FLC dynamic response increases, i.e., the corresponding switching times become shorter. This is especially important for practical applications, because it promotes an increase in the information capacity of FLC displays. If the driving field increases, the FLC response time passes through a minimum, then grows, passes through a maximum at $|\Delta\varepsilon E/4\pi| \propto P_s$ and then decreases again.

For $|\Delta\varepsilon E/4\pi| >> P_s$, the FLC switching times τ are approximately governed by the field squared, $\tau \propto 4\pi\gamma_\varphi/\Delta\varepsilon E^2$, as in the ECB effect in nematic LC (3.10, 3.11). FLC mixtures with negative dielectric anisotropy $\Delta\varepsilon< 0$ are also used for the dielectric stabilization of the initial orientation in FLC displays [26].

As shown above (3.39), response times in the Clark-Lagerwall effect are determined by $\tau_\varphi \propto \gamma_\varphi/P_sE$. The values of $t_{90} - t_{10}$ and $t_{90} - t_{50}$, measured in experiment (Figure 3.34), are fairly close to τ_φ. Comparing the response times in nematics (N):

$$\tau_r^{(N)} \propto 4\pi\gamma_1/ \Delta\varepsilon E^2, \tau_d^{(N)} \propto d^2\gamma_1/K\pi^2 \tag{3.49}$$

and ferroelectric liquid crystals (FLC)

$$\tau_r^{(FLC)} \propto \tau_d^{(FLC)} \propto \tau_\varphi \propto \gamma_\varphi/P_sE, \tag{3.50}$$

we come to conclusion that the electrooptical switching in the Clark-Lagerwall effect in FLCs is much faster than in nematic LC. Slower response of nematic LCs is mainly due to the relatively large decay times $\tau_d^{(N)}$, which in FLC case can be very short in sufficiently high electric fields E.

The optical transmittance I in the Clark-Lagerwall effect is calculated as follows [1-3]:

$$I= \sin^2 4\theta \sin^2\Delta\Phi/2 \tag{3.51}$$

where $\Delta\Phi = 2\pi\Delta nd/\lambda$ the phase difference, $\Delta n = n_{\|} - n_{\perp}$. As follows from (3.51), the maximum contrast is obtained for $0 = \pi/8$ (22.5^0), $\Delta nd/\lambda = 1/2$, which, for $\Delta n = 0.125$, $\lambda= 0.5$ μm, gives d = 2 μm. We should note, that the variation of the cell thickness $\Delta d = \lambda/8\Delta n$ from the optimum value $d = \lambda/2\Delta n$ results in a considerable difference in the FLC electrooptical response (Figure 3.36) [1-3]. The practical criteria of an FLC display quality, however, require more precise limitations d = 2 ± 0.1, μm.

3.5.2.2. Bistable and Multistable Switching in FLC Cells

The bistable switching in SSFLC geometry (Figure 3.35) takes place above a certain threshold field $E_{th} \propto W_d/K^{1/2}$, where K is an average elastic constant and W_d is a dispersion anchoring energy and polar anchoring energy is taken equal to zero W_p=0 [1-3]. Thus with increasing anchoring we have to increase the switching amplitude of the electric field. As the energy of switching electric torque is proportonal to the product of P_sE, the bistability threshold is inversely proportional to the value of the FLC spontaneous polarization P_s.

The problems we meet using the Clark-Lagerwall effect include not only severe restrictions to the optimum layer thickness and requirements of defectless samples, but also difficulties in the realization of a perfect bistability or optical memory switched by the electric field and also providing the grey scale. The latter problem is one of the most crucial, because it is very inconvenient to provide the grey scale using either complicated driving circuits or increasing the number of working elements (pixels) in an FLC display [1-3]. This problem arises in the Clark-Lagerwall effect because the level of transmission is not defined by the amplitude of the driving voltage pulse U, but by the product $U\tau$, where τ is the electric pulse duration.

A surface-stabilized FLC (SSFLC) structure with bistable switching cannot provide an intrinsic continuous gray scale, unless a time- or space-averaging process is applied [1-3]. The inherent physical gray scale of passively addressed FLC cells can be obtained if the FLC possesses multistable electro-optical switching with a sequence of ferroelectric domains, which appear if the spontaneous polarization Ps is high enough [27]. Ferroelectric domains in a helix-free FLC form a quasi-periodic structure with a variable optical density as it appears between crossed polarizers [28] (Figure 3.37). The bookshelf configuration (Figure 3.35) of smectic layers is preferable for the observation of these domains. If the duration of the electric pulse applied to a helix free SSFLC layer containing ferroelectric domains is shorter than the total FLC switching time, the textures shown in Figure 3.37 are memorized after switching off this pulse and short-circuiting of the FLC cell electrodes. The domains appear as a quasi-regular structure of bright and dark stripes parallel to the smectic-layer planes. The bright stripes indicate the spatial regions with a complete switching of the FLC director, while the dark stripes indicate the regions that remain in the initial state. The sharp boundaries between the black and white domain stripes seem to illustrate the fact that only two stable director orientations exist. The variation of the occupied area between bright and dark stripes depends on the energy of the applied driving pulses. The total light transmission of the structure is the result of a spatial averaging over the aperture of the light passed through the FLC cell and is always much larger than the period of the ferroelectric domains. Both the amplitude and the duration of the driving pulses can be varied to change the switching energy, which defines the memorized level of FLC-cell transmission in a multistable electro-optical response. Therefore, any level of the FLC-cell transmission, intermediate between the maximum and the minimum transmissions, can be memorized after switching off the voltage pulses and short-circuiting of the cell electrodes (Figure 3.37).

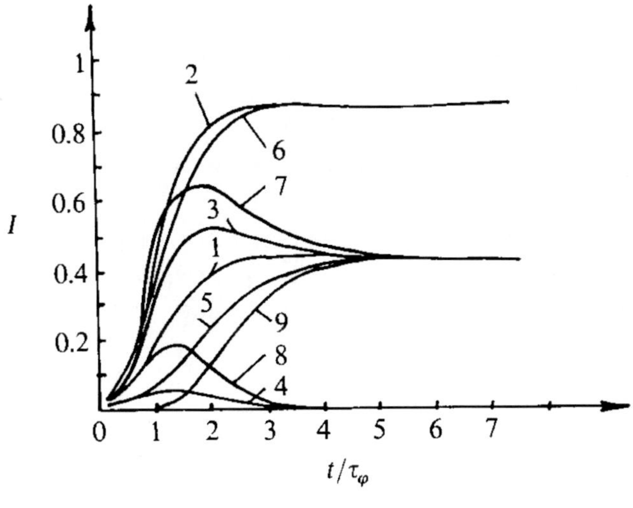

Figure 3.36. Electrooptical response of the FLC cell in Clark-Lagerwall effect for different phase factors $d\Delta n/\lambda = k/4$, k= 1…9 [1-3].

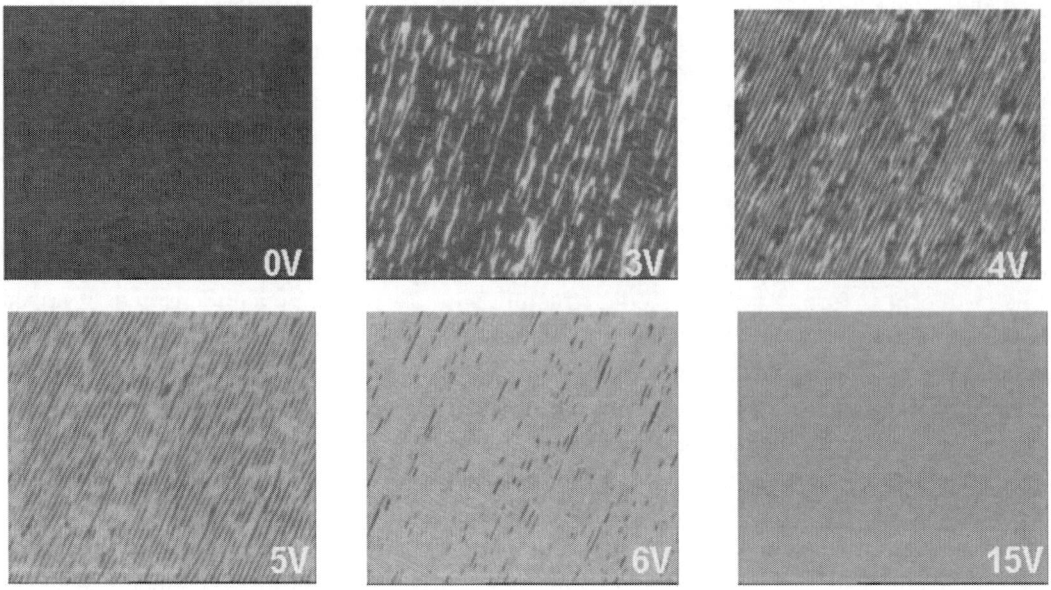

Figure 3.37. Continuous variation of the width of ferroelectric domains with a change in the applied voltage of the FLC layer between crossed polarizers [28].

3.5.2.3. Optical Transmission of Multistable FLC Cell

The necessary conditions of multistable switching modes, are (i) sufficiently high FLC spontaneous polarization Ps > 50 nC/cm2 and (ii) a relatively low energy of the boundaries between the two FLC states existing in FLC domains (Figure 3.37), which is usually typical for the antiferrroelectric phase [1-3].

The multistability is responsible for three new electro-optical modes with different shapes of the gray-scale curve that can be either S-shaped (double or single dependent upon the applied-voltage pulse sequence and boundary conditions) or V-shaped dependent upon boundary conditions and FLC cell parameters [28] (Figure 3.38).

The simple approach describes well the experimental curves of average intensity of multi-stable FLC states (S and V-shape) for various position of the FLC texture between crossed polarizers. Consider a homogeneous smectic C* layer shown in Figure 3.39. The analyzer is crossed to the polarizer. The normal of the smectic C* layer (z-axis) is placed at the angle β to the polarizer.

The director **L** of liquid crystal rotates around the z-axis with the cone angle β, which can be described as follows by Eq. (3.37), where elastic terms have been neglected since the applied field E and FLC spontaneous polarization P_s are sufficiently large in our case.

Eq. (3.37) can be solved as

$$\varphi(t) = 2arctg\left(A \cdot \exp\left(\frac{E\tau_p P_s}{\gamma_\varphi}\right)\right) \tag{3.52}$$

where $A = tg\left(\frac{\varphi_0}{2}\right)$ with φ_o (φ $(t=0)$) is the initial phase angle, τ_p is the pulse duration of driving rectangular voltage. The relative light transmission $I=J/J_0$ (where J and J_0 is an FLC cell input and output intensity) of the smectic C* layer placed between crossed polarizers (Figure 3.39), can be written as [27]:

$$I = \sin^2 2(\beta - \theta_0 \cos\varphi)\sin^2 \frac{\pi \Delta n d_{FLC}}{\lambda}, \tag{3.53}$$

where λ is the wavelength, Δn is the birefringence index, d_{FLC} is the FLC layer thickness, β is an angle between the polarizer plane and a normal to smectic layers as it is shown in Figure 3.39. According to (3.53) the necessary condition of a maximal light transmission is:

$$\frac{\Delta n d_{FLC}}{\lambda} = \frac{1}{2}, \frac{3}{2}... \tag{3.54}$$

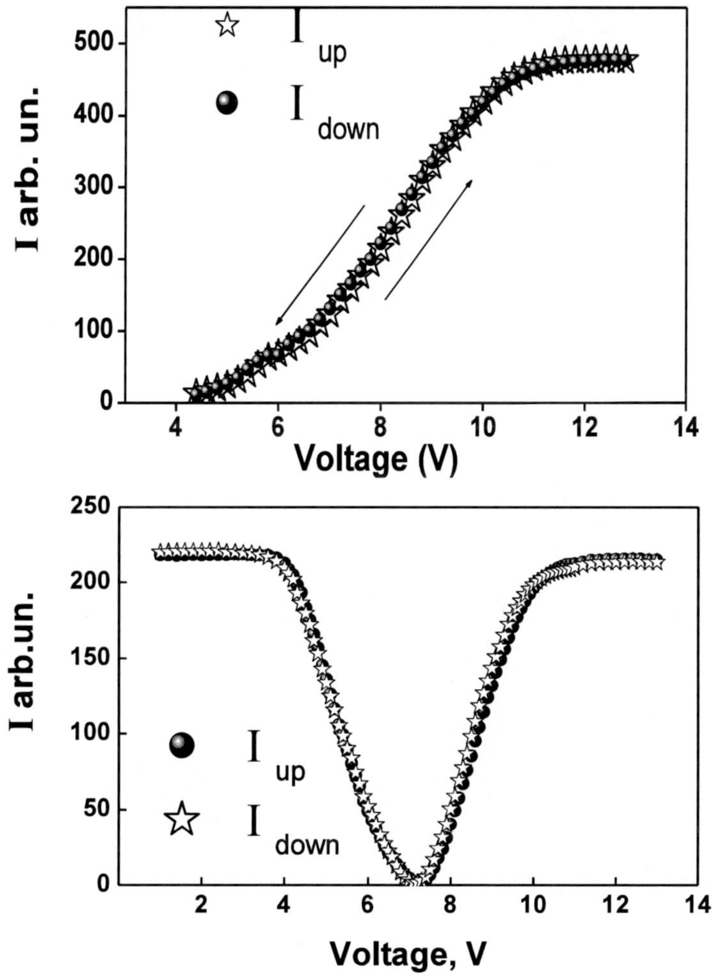

Figure 3.38. S-shaped (above) and V-shape (below) FLC multi-stable switching [28].

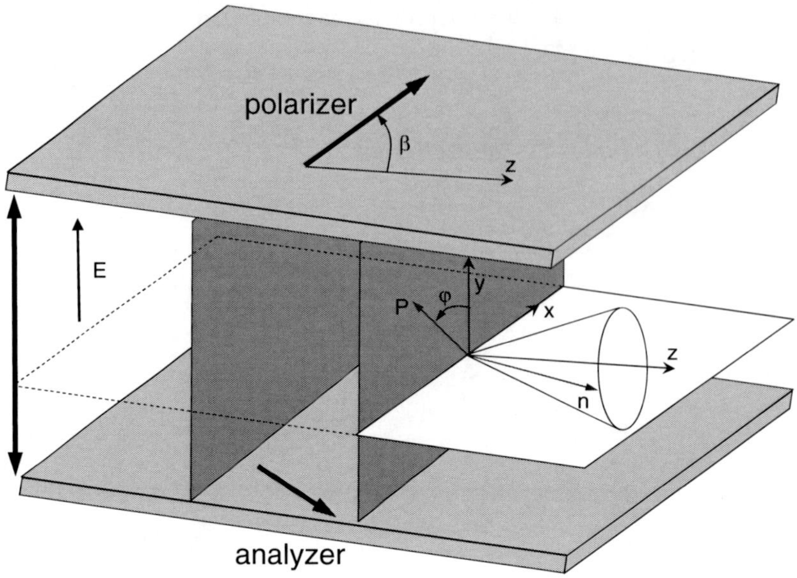

Figure 3.39. Multistable FLC cell configuration.

Taking into account (3.54) as a necessary condition we consider two cases:

$$\beta = \theta_0 \tag{3.55}$$

and

$$\beta = 0 \tag{3.56}$$

The memorized transmission level I evaluated according to (3.52)-(3.53) writes:

$$I = \sin^2(4\theta_0 \frac{f^2}{1+f^2})\sin^2(\frac{\Delta\Phi_0}{2} - 2\delta_0 \frac{f^2}{(1+f^2)^2}), \tag{3.57}$$

where

$$f = A \cdot \exp\left(\frac{E\tau_p P_s}{\gamma_\varphi}\right), \; f >> 1 \tag{3.58}$$

Further, in (3.57) we have:

$$\delta_o = \frac{\pi d n_{//}}{\lambda}(\frac{n_{//}^2}{n_\perp^2} - 1)\theta_o^2, \tag{3.59}$$

where $n_{||}$ and $n_\perp$ are refractive indices, parallel and perpendicular to long molecular axis correspondingly, and

$$\Delta\Phi_o = \frac{2\pi d}{\lambda}(n_{//} - n_{\perp}) \tag{3.60}$$

If conditions (3.54) and (3.56) are satisfied then the memorized FLC cell transmission level *I* evaluated according to (3.53) can be described as:

$$I = \sin^2(2\theta_0 \frac{f^2 - 1}{f^2 + 1})\sin^2(\frac{\Delta\Phi_0}{2} - 2\delta_0 \frac{f^2}{(1 + f^2)^2}) \tag{3.61}$$

Numerical evaluations according to our simple model show that *I(E)* dependence calculated from (3.57) is S-shaped, while the evaluations from (3.61) give V-shaped mode (Figure 3.38).

Thus the theoretical and experimental investigation of reversible and memorized S and V-shaped multistable FLC electro-optical modes was proposed [27]. New electro-optical modes are based on the multistable electro-optical modes in FLC cell (Figure 3.37).

3.5.2.4. Deformed Helix Ferroelectric Effect

The geometry of the FLC cell with a DHF effect is presented in Figure 3.40 [29]. The polarizer (P) on the first substrate makes an angle with the helix axis and the analyzer (A) is crossed with the polarizer.

The FLC layers are perpendicular to the substrates and the layer thickness d is much higher than the value of the helix pitch R_o:

$$d >> R_o. \tag{3.62}$$

The light beam with the aperture a >> R_o passes parallel to the FLC layers through an FLC cell placed between the polarizer and analyzer. In an electrical field the FLC helical structure becomes deformed, so that the corresponding dependence of the director distribution cosφ, as a function of coordinate $2\pi z/R_o$, oscillates symmetrically in ±E electric fields (Figure 3.40).

These oscillations result in a variation of the effective refractive index, i.e., electrically controlled birefringence appears.

The effect takes place up to the fields of FLC helix unwinding

$$E_u = \pi^2/16\ K_{22}q_0^2/P_s, \tag{3.63}$$

where K_{22} is FLC twist elastic constant, $q_0 = 2\pi/R_0$ –helix wave vector.

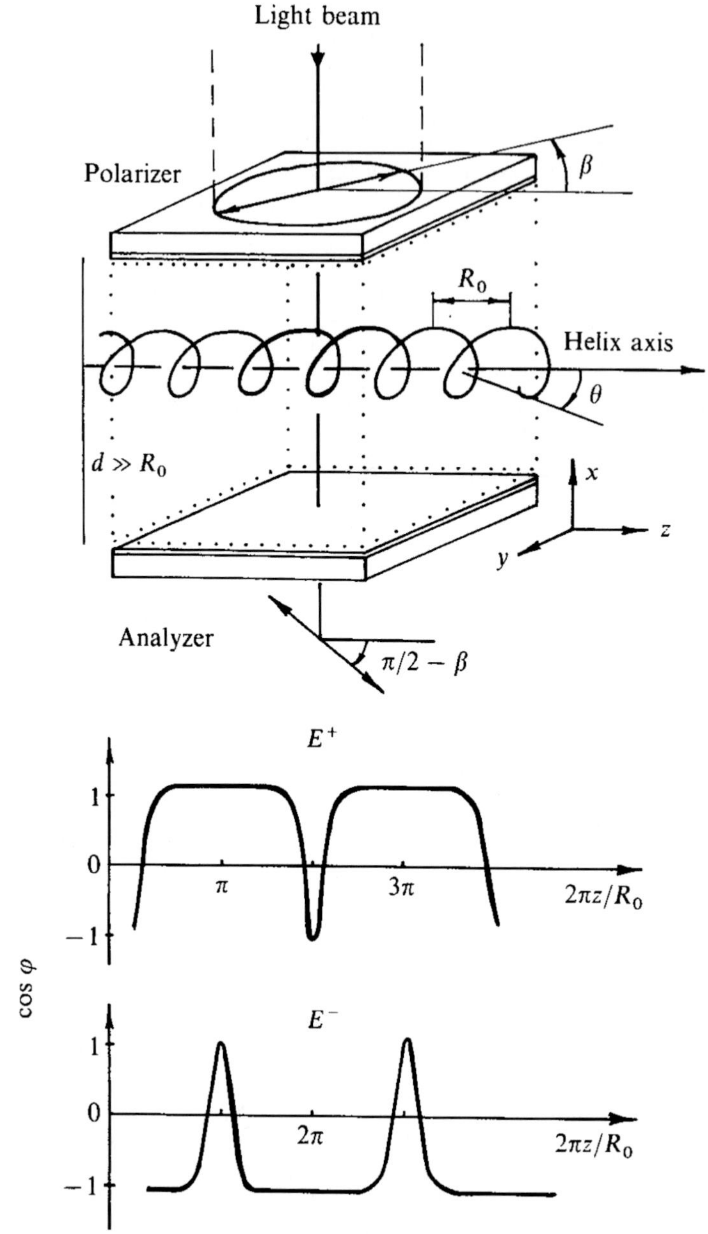

Figure 3.40. Deformed Helix Ferroelectric (DHF) Effect [29].

The characteristic response times τ_c of the effect in small fields $E/E_u << 1$ are independent of the FLC polarization P_s and the field E, and defined only by the rotational viscosity γ_φ, and the helix pitch $R_o = 2\pi/q_0$:

$$\tau_c = \gamma_\varphi/K_{22}q_0^2 \tag{3.64}$$

The dependence (3.64) is valid, however, only for very small fields E. If $E \le E_u$, the FLC helix becomes strongly deformed and $\tau_c \propto E^{-\delta}$, where $0 < \delta < 1$ [30]. If E is close to the unwinding field E_u the helical pitch R sharply increases $R >> R_o$. Consequently, the times of the helix relaxation τ_d to the initial state also rise

$$\tau_d / \tau_c \propto R^2/R_0^2 \tag{3.65}$$

i.e., for $E \approx E_u$ it is possible to observe the memory state of the FLC structure [29]. In this regime the electrooptical switching in the DHF effect reveals a pronounced hysteresis especially for $E \Rightarrow E_u$.

However, if the FLC helix is not deformed too much, fast and reversible switching in the DHF mode could be obtained [1-3]. The switching time, less than 10 μs at the controlling voltage of ±20V can be provided, which is temperature independent over the broad temperature range [31]. The fast ferroelectric LC cells (deformed helix ferroelectric-DHF effect) with the response time less than one microsecond in a broad temperature range from 20 to 80 degrees were developed [32] (Figure 3.41). We believe, that DHF FLC are among the fastest electrooptical mode in LC cells for photonics and display applications.

The optical transmission of the DHF cell could be calculated as follows:

$$I= \sin^2 (\pi \Delta n(z)d/\lambda)\sin^2 (2(\beta-\alpha(z))) \tag{3.66}$$

where β is the angle between the z-axis and the polarizer (Figure 3.40),

$$a(z) = \arctan(\tan \theta \cos \varphi(z)) \tag{3.67}$$

is the angle between the projection of the optical axis on the y, z-polarizer plane and the z-axis, and $\Delta n(z) = n_{eff}(z) - n_\perp$ is the effective birefringence:

$$n_{eff} = n_\| n_\perp /[n_\perp^2 +(n_\|^2 - n_\perp^2) \sin^2\theta \sin^2\varphi]^{1/2} \tag{3.68}$$

(Diffraction of light on the helical structure is avoided, because the helical pitch of the FLC mixture $R_o \propto 0.1\text{-}0.3$ μm < λ =0.5 μm is in the visible range).

In the case of small angles $|\theta| << 1$, the transmission in (3.66) can be expanded in θ series

$$I \propto (\sin^2 2\beta - 2\theta \sin 4\beta \cos\varphi + 4\theta^2 \cos 4\beta \cos^2\varphi)\sin^2 \pi \Delta n d/\lambda \tag{3.69}$$

As shown in [1-3] for the small values of the applied field $\cos \varphi \propto E/E_u$ and changes its sign for the field reversal $E \Rightarrow -E$ (Figure 3.40). Thus according to (3.69) for sin4β=0 we have a quadratic grey scale, i.e.

$$\Delta I \propto \theta^2 \cos^2\varphi \propto \theta^2 E^2/E_u^2 \tag{3.70}$$

and for other values of θ the grey scale is linear. For cos4β=0 the quadratic component in the modulated intensity I is absent, i.e.

$$\Delta I \propto \theta\cos\varphi \propto \theta\, E/E_u \tag{3.71}$$

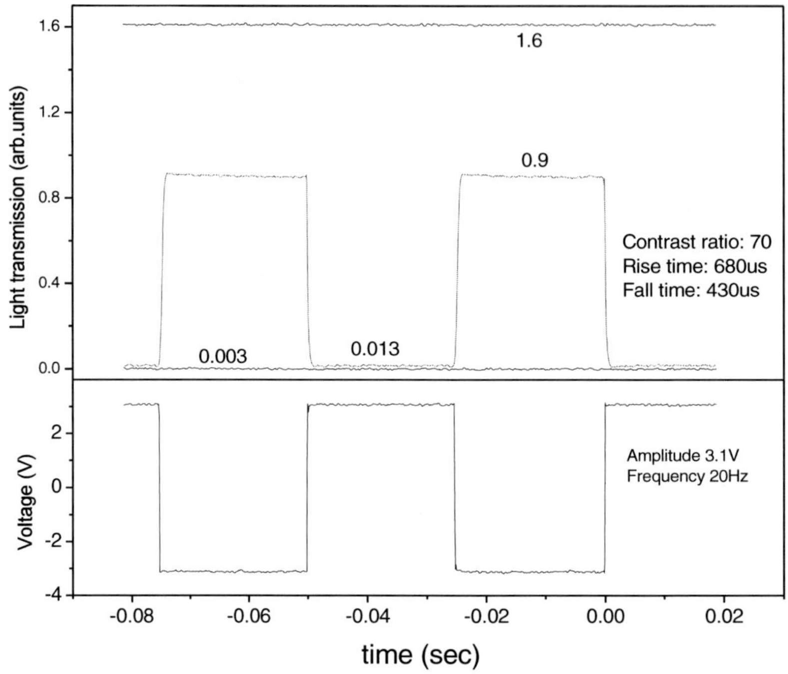

Figure 3.41. Switching time of deformed heix ferroelectric (DHF) LC cells [32].

If E(t) = E_o coswt, then in the case of (3.70) we come to the modulation regime, which doubles the frequency of the applied field. The relationships (3.70), (3.71) were confirmed in experiment [29]. Using a "natural" grey scale of the DHF mode many grey levels have been obtained with fast switching between them [33]. New ferroelectric mixtures with the helix pitch $R_0 < 0.3\mu m$ um and tilt angle $0 > 30^0$ have recently been developed for the DHF effect [33-35]. The helix unwinding voltage was about 2-3 V. Short-pitch FLC mixtures could be also used to obtain pseudo-bistable switching in FLC samples. Using these new FLC's, the electrically controlled V-shaped switching in DHF mode can be applied for new active-matrix LCD with Field Sequential Colors (FSC).

A geometry with $\beta=0$ was selected in all experiments to provide non-sensitive to the driving voltage polarity electro-optical response. Maximal light transmission under this condition, as it follows from (3.53), occurs if $\alpha(z) = 45^0$ and $\Delta n(z)d=\lambda/2$. It is easy to show from (3.53) that the tilt angle θ of the FLC should be close to 45° for providing of maximal light transmission at $\beta=0$. The typical V-shape symmetrical (voltage sign independent) DHF switching is shown in Figure 3.42 [33].

Let us point to certain advantages of the DHF electrooptical effect for applications as compared with the Clark-Lagerwall mode [1-3].

1. High operation speed is achieved for low driving voltages. This takes place because a slight distortion of the helix near the equilibrium state results in a considerable change in the transmission. Consequently, the instantaneous response of the FLC cell is provided without the so-called delay time inherent to the Clark-Lagerwall effect.
2. The DHF effect is also less sensitive to the surface treatment and more tolerant to the cell gap inhomogeneity. As follows from experiment and qualitative estimations [1-3] the effective birefringence value Δn_{eff} is approximately twice as low as $\Delta n = n_{||} - n_{\perp}$ in the Clark-Lagerwall effect.
3. The DHF effect allows the implementation of a "natural," i.e., dependent on voltage amplitude, grey scale both linear and quadratic in voltage. Moreover, at $E \approx E_u$ long-term optical memory states are possible.

In particular, a high quality dark state of DHF FLCD has been obtained as well as V-shape switching with a high frequency [33] (Figure 3.42). Two inherent properties of DHF-FLC have been observed (Figure 3.42): (i) similar to NLC cells, the electro-optical response is insensitive to the driving voltage polarity, but the response time is two orders faster as compared with NLC, i.e. $\tau_{0.1-0.9}^{on} \cong 80\mu s$, $\tau_{0.1-0.9}^{off} \cong 60\mu s$; (ii) the response exhibits perfect V-shaped mode obtained for the case of rectangular alternating applied voltage pulses shown in Figure 3.42. As is illustrated in Figure 3.42, the electro-optical response frequency is twice higher than the driving voltage frequency. The cell was controlled by the voltage waveform appropriate for FSC displays.

Figure 3.42 shows that the cell light transmission is continuously tunable for each color field independently at both voltage polarities.

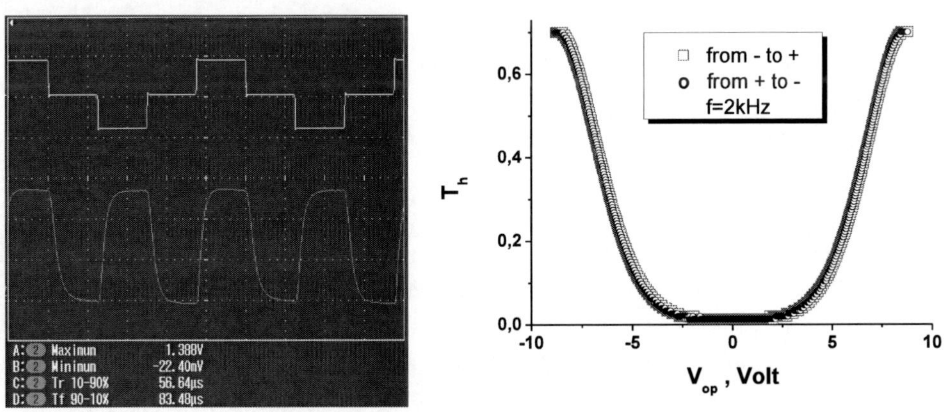

Figure 3.42. Symmetric (voltage sign independent) electrooptic response of DHF –FLC [33]. Left: Top – the driving voltage waveform applied to the cell, bottom – the electro-optical response of the cell. Right: V-shaped mode in the envelope curve of light transmission saturation states measured at electro-optical response frequency 2 kHz. The light transmission Th evaluated in comparison with transmission of empty cell placed between parallel polarizers; this transmission is defined as Th = 1.

Thus V-shaped electro-optical response is shown to be an inherent property of a DHFLC cell under a special choice of the applied rectangular alternating electric field waveform, frequency, and the cell geometry. In contrast to other known V-shaped FLC modes, the discovered V-shaped switching is observed in a broadband frequency range including one kHz, and not at a certain characteristic frequency. Frequency independent V-shaped DHFLC switching allows increasing drastically (up to 1 kHz) the operating frequency of FSC LCD cells.

A special software for an advanced design of new FLCD prototypes [36,37]. Certain new FLCD prototypes can be optimized using our software, including the optimization of FLC configuration as well as the optical scheme.

One of remarkable properties of DHF mode is independence of the response time on applied voltage at V<Vc. Another characteristic feature of DHF mode is frequency dependence of the electro-optical response time, which is really exists in our case shown in Figure 3.43 We have found that the switching time of DHF FLC decreases with frequency of the applied voltage, which is in a good agreement with our experimental results (Figure 3.43).

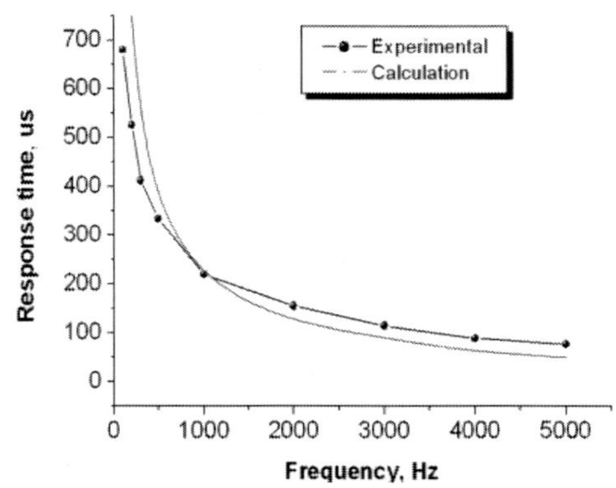

Figure 3.43. Response time of DHF FLC as a function of the applied voltage frequency.

3.5.2.5. Electrically Suppressed Helix (ESH) Mode

A new electrooptic mode, found recently was called Electrically Suppressed Helix (ESH) mode [35,38]. The mode is characterized by a high contrast ratio and very fast electrooptic response and is highly suitable for display and photonics applications. Deformed helix ferroelectric (DHF) mode exists at $V \leq V_c = 0.4$V and the helix is completely suppressed by electric field at V≥1V, see Figure 3.44 right part and linear dependence of the inverse response time at V>1V in Figure 3.45, where the response time has absolutely the same behavior as in SSFLC mode [38]. Despite of these similarities, it should be emphasized that at $V > V_c$, strictly speaking, there is no SSFLC mode but an electro-optical mode with completely electrically suppressed helix arises (ESH-mode). It would seem that between SSFLC and ESH modes there is only a formal difference in initial conditions (the helix is suppressed by surfaces in SSFLC, but this does not happen with ESH), as the dynamics of a completely identical. Actually, the presence of helix without applied voltage is the cause of the unique high alignment quality in ESH-mode (see top right insertions in Figure 3.44). The contrast ratio in ESH mode is more than 12000:1 up to 1 kHz at ±1.5V and up to 2 kHz at ±3V; the light transmission is about of maximum in this case, see insertion to Figure 3.46.

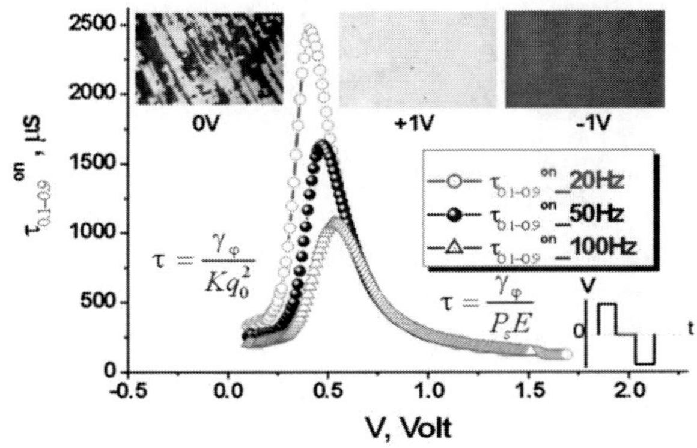

Figure 3.44. Dependence of switching on time of 1.5 μm cell filled with the FLC-595 [38]. *Insertions*: bottom right is the driving voltage waveform; top the FLC layer textures between crossed polarizers, horizontal scale of all images is 200 μm; top left - the voltage is not applied, top right - textures at +1V and -1V.

The unique high alignment quality, however, is observed only if the helix elastic energy is comparable but obligatory not less than the anchoring energy normalized to d_{FLC}:

$$Kq_0^2 \geq \frac{2W_Q^0}{d_{FLC}} \tag{3.72}$$

where W_Q^0 is a coefficient the anchoring energy. We measured W_Q^0 = 4·10^{-4} J/m^2 just for the cell whose properties are illustrated in Figures 3.44, 3.45, thus confirming the validity of inequality (3.72), when outstanding alignment quality (top right insertions in Figure 3.44) is observed.

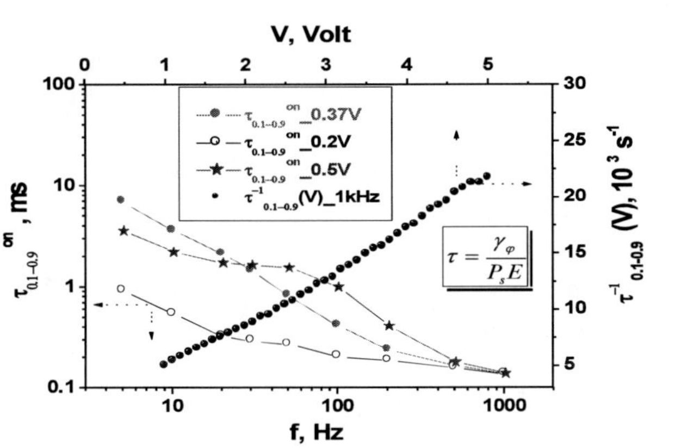

Figure 3.45. The frequency dependencies of the response time at different voltages under condition V≤0.5V; (•) -linear dependence of the inverse response time $\tau^{-1}_{0.1-0.9}(V)$ at V>1 V [38].

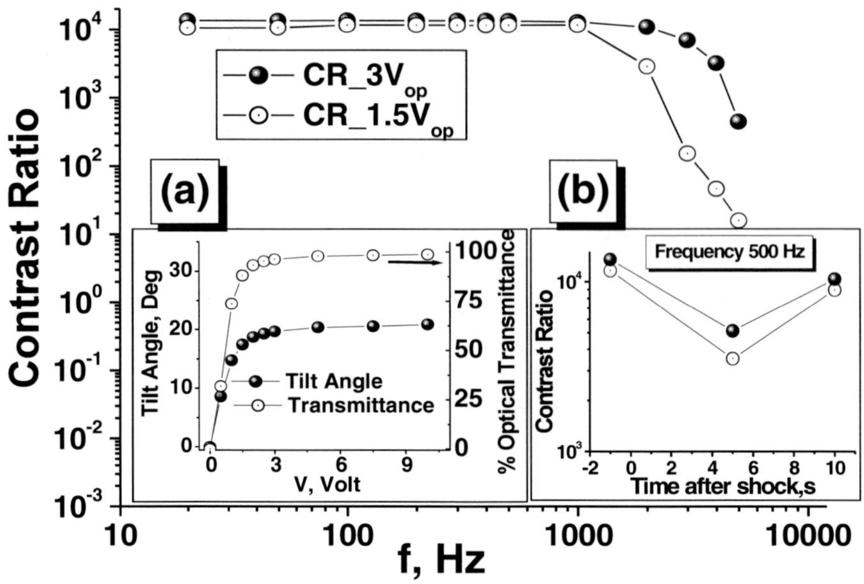

Figure 3.46. Dependencies of the contrast ratio on driving voltage frequency at V=1.5V and V=3V. The FLC-595 [38] layer thickness is 1.5μm, no SiO_2 layers, measurements have been carried out at T=22°C, wavelength λ=0.63 μm. Insertion left: Dependencies on voltage: measured tilt angle and the cell light transmission in comparison with transmission of empty cell placed between crossed polarizers. Insertion right: Contrast ratio dependence on time after the application of mechanical shock.

A very uniform FLC alignment exists just only in ESH-mode when $V > V_c$ (right top insertion in Figure 3.44) while at $V<V_c$ it is worse because the FLC layer is divided into two-domain helical structures with the domains principal axes deployed at an angle around double FLC tilt angle 2θ. When observed with a polarizing microscope, these helical domains are perceived as homogeneous dark and light areas with their characteristic sizes around 10-50 μm, see left top insertion in Figure 3.44. The domains are completely suppressed by a weak electric field $E_c<E<1V/\mu m$ (right top insertion in Figure 3.44) thus forming perfect chevron-free alignment in ESH-mode.

In addition to this the shock stability has also been found good for the ESH mode. The contrast ratio was measured with time after removing the mechanical pressure. The contrast can be restored in a very short time after removal of mechanical pressure (Figure 3.46). The electro-optical response manifests evident saturation at 500 Hz and ±1.5V with the response time $\tau_{0.1\text{-}0.9}$ =130 μs (Figures 3.44, 3.45).

3.5.2.6. Orientational "Kerr Effect" in DHF FLC with Subwavelength Pitch

The Kerr like non-linearity has been observed in ferroelectric liquid crystals (FLC) with sub-wavelength helix pitch [39,40]. In order to systematically examine the pure phase modulation of light that takes place when the selective reflection from the helix is suppressed, FLC with the helix pitch R_0 = 150 nm was used. Vertical alignment of the FLC with the well-known geometry of inter-digital electrodes have been applied to orient the helix axis perpendicular to the substrates while electric field is parallel to smectic layers. In the absence of electric field, the FLC helical structure was optically equivalent to a uniaxial crystal whose principle axis coincides with the axis of the helix. In the presence of electric field E, the in-plane refractive indices along (n_{II}) and perpendicular ($n_{\square}$) to the field differ from each other and their difference was proportional to square of the field: $\delta n_{ind} \infty E^2$. Thus, proposed phase modulator exhibits the hysteresis free 2π phase modulation with constant ellipticity of emerging light and response time less than 100 μs. The described Kerr effect could find application in many modern electro-optical devices.

Figure 3.47 (a) presents the extensive overview of the electro-optical cell with standing FLC helix and inter-digital electrode deployed on one of the two glass plates [39,40]. The inter-digital electrodes with the electrode width of 2μm and electrode gap of 50 μm have been deployed to apply electric field parallel to the smectic layers. The vertical alignment to the DHF FLC has been accomplished by spin coating glass plates with 40 nm thick layer of stearic acid chromium salt (Figure 3.47 (a)) followed by soft backing at 100 °C for 10 min. The Figure 3.47 (c) represents the optical micro-photograph of the a vertically aligned DHF FLC (VADHFLC) cell under crossed polarizers and in the absence of the electric field.

Figure 3.48 (a) reveals the schematic diagram of the VADHFLC cell where the DHFLC principal axis (PA) and its interaction with the applied electric field. In the presence of electric field E, (i.e. $E < E_c$), the PA deviates from its initial position z by an angle $\Delta\alpha$ in a plane perpendicular to the electric field direction and as a consequence the refractive indices of the helical structure along ($n_{E_{II}}$) and perpendicular ($n_{E_{\perp}}$) to the electric field (Figure 3,48 b) differ from each other and their difference corresponds to the biaxiality (δn_{ind}) of the helical structure [39,40].

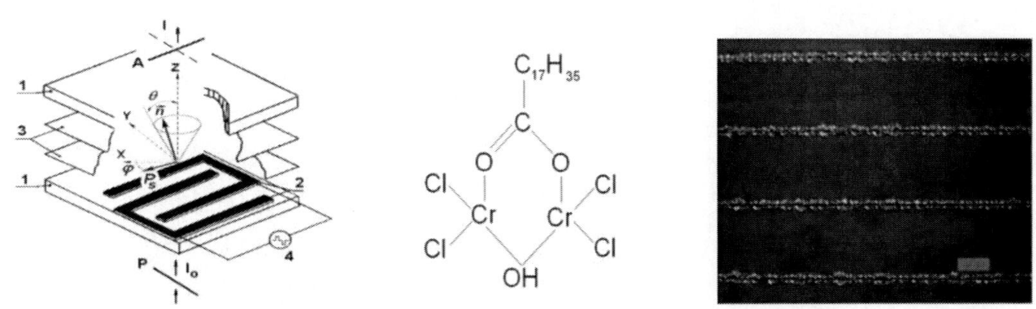

(a) (b) (c)

Figure 3.47. Schematic representation of geometry of a vertically aligned DHFLC cell: 1 are glassy plates, 2 are in-plane electrodes, 3 are smectic layers parallel to glassy substrates, 4 is a voltage generator, P (A) is polarizer (analyzer), θ is the tilt angle of director $\boldsymbol{L}$ in smectic layers, φ is azimuthal angle of the spontaneous polarization vector $\boldsymbol{P}_s$, I_0 is intensity of incident light, I is intensity of light transmitted through the VADHFLC cell placed between P and A. (b) the chemical structure of stearic acid chromium salt, (c) the micro-photograph of VADHFLC cell placed between crossed polarizers at V=0. The red mark is equal to the 20μm [39,40].

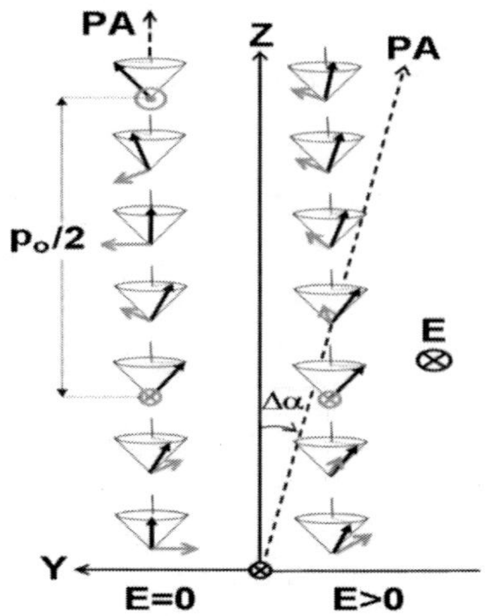

(a)

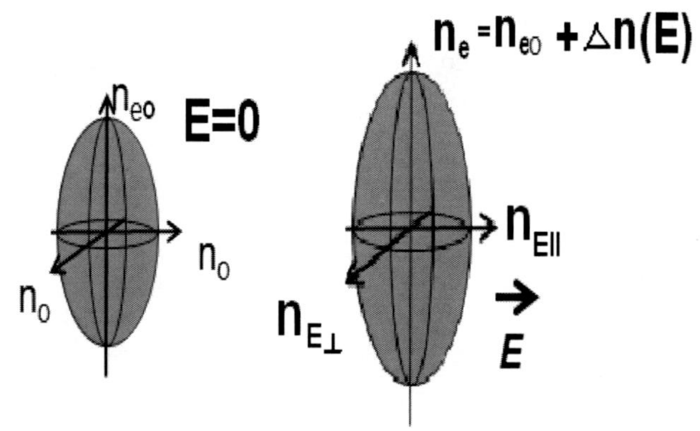

(b)

Figure 3.48 (a) illustration of the VADHFLC principal axis deflection in the presence of electric field, (b) schematic representation of the refractive indices ellipsoid in the absence (left) and in the presence (right) of electric field [39,40].

For normally incident light, the electrically controlled birefringence writes [39,40]:

$$\delta n_{ind} = n_{\|E} - n_{\perp E} = K_{Kerr} E^2, \tag{3.73}$$

where $n_{\|E}$ and $n_{\perp E}$ the effective DHF FLC refractive indexes, induced parallel and perpendicular to the appluied electric field and "Kerr constant"

$$K_{Kerr} = 2\, n_{E=0} \cdot (1 - n^2_{\perp} / n^2_{E=0} \cdot \chi_E / P_s) \tag{3.74}$$

is proportional to the the "isotropic" refractive index for zero electric field $\alpha_{Kerr} \sim n_{E=0}$, shown below:

$$n^2_{E=0} = [n^2_{\|} n^2_{\perp} / (n^2_{\perp} + (n_{\|}^2 - n_{\perp}^2) \cos^2\theta)^{1/2} + n^2_{\perp}]/2,$$

$n_{\|}$ and $n_{\perp}$ FLC refractive indexes and θ- FLC tilt angle, P_s-FLC spontaneous polarization, and $\chi_E = \partial P_s < \cos\varphi > / \partial E$ - FLC average dielectric susceptibility along the field direction.

The FLC transmittance T as a function of the E^2 is shown in Figure 3.49 [39,40]. This figure reveals that the VADHFLC cell shows the 2π phase modulation at the electric field of ~ 2.1 V/μm. The light transmission of VADHFLC cell between two crossed polarized can be described by [39,40]:

$$T = \sin^2 2\Psi \sin^2 \frac{\pi \delta n_{ind} d_{FLC}}{\lambda}, \tag{3.75}$$

where d_{FLC} corresponds to the FLC layer thickness. Thus, the δn_{ind} has been computed from Eqs. (3.73) and plotted in Figure 3.49. The open circular legends in Figure 3.49 represent the experimental data while the solid blue line represents the theoretical fit of Eqs. (3.73). The value of δn_{ind} for FLC-587 at saturation is 0.05, which means that 2π modulation can be obtained for the FLC thickness to wavelength ratio $d_{FLC}/\lambda=20$ (Fig 3.49). This is a reasonable thickness of FLC cell, which will still keep the high quality FLC orientation even for infrared (IR) light [39,40].

With the help of experimental results for δn_{ind} and eq. 3.73, K_{kerr} has been evaluated at different frequencies and plotted in the insertion of Fig 3.49. The K_{kerr} = 27nm/V^2, which is about two orders of magnitude greater than the Kerr constant of nitrobenzene and twice larger than the Kerr constant of the best known polymer stabilized blue phase LC (PSBPLC) [41]. The K_{kerr} is constant till the frequency of 2 kHz afterwards, at higher frequencies, it shows clear dispersion. This dispersion can also be explained on the basis of eq. 3.73. The K_{kerr} depends on the dielectric susceptibility of the goldstone mode (χ_E) that is defined as [39,40]:

$$\chi_E = \frac{P_s^2}{2Kq_0^2} \tag{3.76}$$

where K is the elastic constant while the $q_o = 2\pi/R_0$ is the wave vector of the helix, thus K_{kerr} can be increased further by optimizing the FLC material parameters.

The phase modulation of light in FLC orientational "Kerr effect" occurs only because of the electrically induced variation of the FLC refractive indices. Consequently, no modulation in the ellipticity of output light has been observed (Figure 3.50) [39,40]. For $E<E_c$, the ellipticity is independent on the electric field which is the same range of the electric field where the δn_{ind} remains quadratic. However, near E_c and at higher electric field the $\delta n_{ind}(E)$ deviates from the quadratic dependence on E and simultaneously the ellipticity of the emerging light begins to increase. This increase in the ellipticity is because of the helix unwinding process near E_c that results in polarization plane rotation of impinging light.

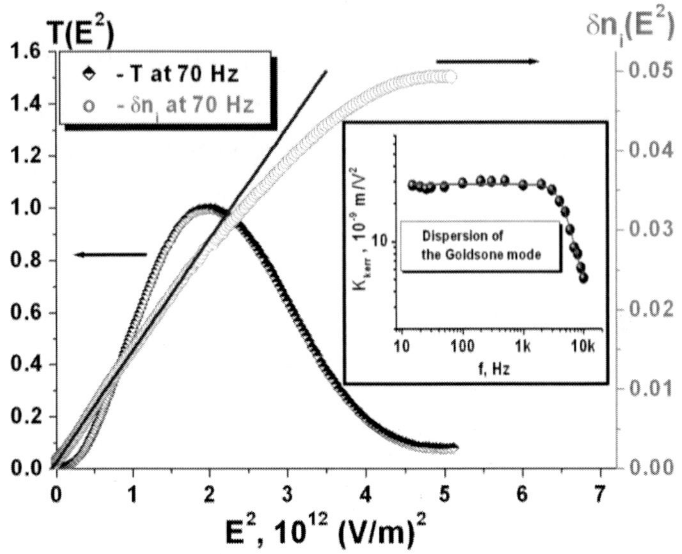

Figure 3.49. The transmittance $T(E^2)$ (♦) and $\delta n_{ind}(E^2)$ (○) dependence for VADHFLC cell placed between crossed P and A at $\Psi = 45^\circ$ at the applied frequency of 70 Hz. The solid blue line is the result of theoretical fit of the δn_{ind} according to eq. (3.73). Insertion: shows the plot of Kerr constant "K" with frequency and solid red line represent the theoretical fit [39,40].

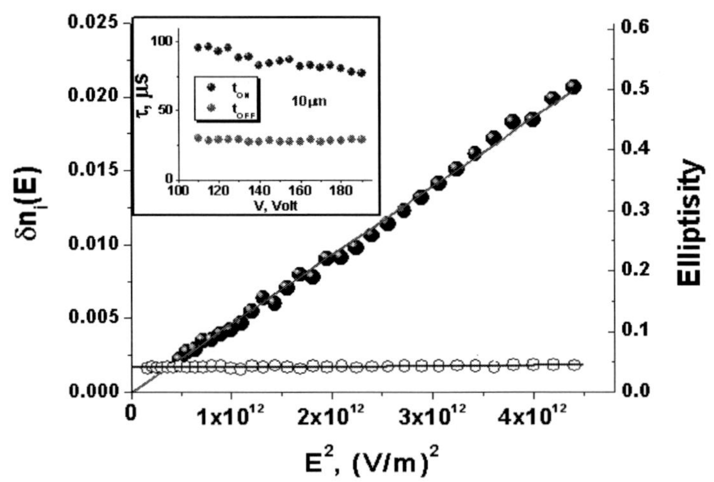

Figure 3.50. Solid legends represents the δn_{ind} of 18μm thick vertically aligned DHFLC cell and open legends are for the ellipticity of the emerging light at the frequency of 40Hz. The insertion represents variation of the t_{ON} and t_{OFF} with respect to the applied voltage [39,40].

The insertion in Figure 3.50 represents the electric filed dependence of the t_{ON} and t_{OFF}. The VADHFLC device is characterized by the fast response time less than 100 μs and thus the modulation frequency could be as high as 1 kHz with saturated electro-optical states. In recent time PSBPLC has been considered as emerging display and phase modulating technology however, the VADHFLC seems more appropriate because of higher Kerr constant, hysteresis free electro-optics and faster (around an order of magnitude) electro optical response . Thus, VADHFLC could find application in many modern electro-optical devices.

3.6. SURFACE BISTABILITY IN NEMATIC LC

We have described above the elerctrooptical effects in cholesteric and ferroelectric LC as well as bistable and multistable cholesteric and ferroelectric LCD with a low power consumption and memory. We are going to provide some more examples of these low power consumption LCD below based on surface bistability in nematic LC.

3.6.1. BTN with 0⇔π Twist Angle Switching

A bistable TN (BTN) with 0⇔π twist angle switching bistable device using monostable surface anchoring switching was demonstrated [42]. The two bistable textures were planar and π twisted alignments (Figure 3.51). The switching to one or another bistable state from the initially homeotropic configuration, obtained in a strong electric field depends on the rate of the field decrease and associates with an anchoring breaking on the "slave" substrate (Figure 3.51). Slow decreasing field down to zero results in a planar state, while turning off the field instantaneously initiates a strong back flow effect thus forces the transformation to π twisted state.

The cell gap of BTN LCD (d ~ 1.5 μm) is thinner than the one of conventional LCDs. The top and bottom substrates are are rubbed in the same direction (Figure 3.51). One substrate ("slave" substrate) is coated with a special alignment layer, which gives a nearly planar anchoring with moderately strong polar anchoring energy ($W_\theta \approx 3{\cdot}10^{-4}$ J/m^2). The other substrate ("master" substrate) is coated with a conventional polyimide, which gives tilted anchoring (~5°) with a strong polar anchoring energy ($W_\theta \approx 10^{-3}$ J/m^2). Both azimuthal anchoring energies are kept strong enough ($W_\varphi \approx 10^{-4}$ J/m^2) to maintain the azimuthal orientation of the liquid crystal on the alignment layer. Two stable textures U and T exist. The U texture is 0°-twisted while the T texture is 180°(π) twisted state. A chiral dopant is added to the liquid crystal mixture to adjust the free pitch about four times higher than the LC cell thickness to equalize the energies of U and T. U and T appear respectively black and white in the standard reflective configuration [42]. A substantial problem of BTN LCD is to find a proper surface with a moderate polar, but still large azimithal anchoring energy, which is not easy. Typical BTN LCDs are shown in Figure 3.52.

A novel switching mechanism for bistable nematic twisted displays (BTN) was proposed [43]. By using the bi-directional alignment surface, which is fabricated by a photo-polymerizable polymer, 0⇔π twist angle switching can be controlled by changing the voltage levels. Such switching avoids any anchoring breaking, hence large cell gap can be possible. Furthermore, since the switching depends on the voltage level, it is completely compatible with ordinary low cost STN drivers [43].

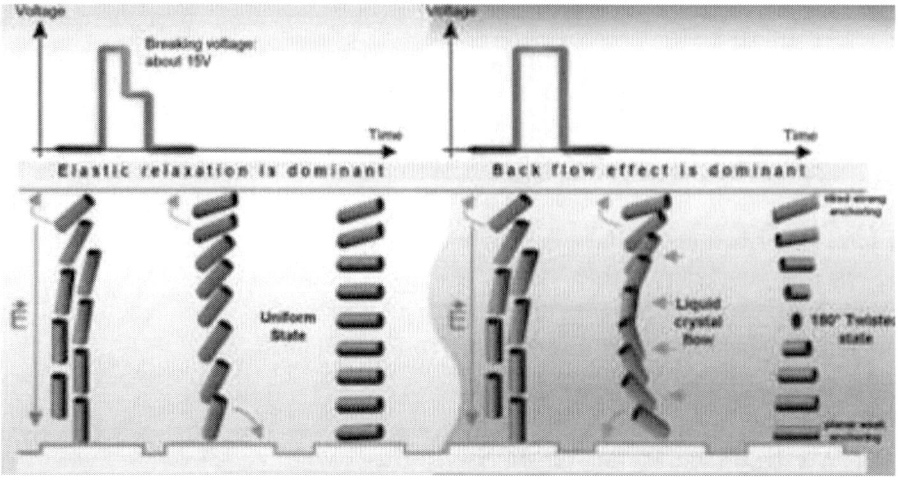

Figure 3.51. Bistability in nematic LC [42]. Slow decreasing field down to zero results in a planar state, while turning off the field instantaneously forces the creation of a bend state, which spontaneously transforms to π twisted state.

Figure 3.52. Typical BTN LCD (courtesy by Nemoptic).

3.6.2. Zenithal Bistability (ZBD)

A grating surface treated with a homeotropic surfactant was found to produce two bistable pretilt LC configurations, when the grooves depth to pitch ratio is an optimum [44]. Bistability results for a range of grating shapes between a stable shallow grating, and a stable deep grating [44]. Current production devices use a grating with 0.8μm pitch and 0.9μm amplitude.

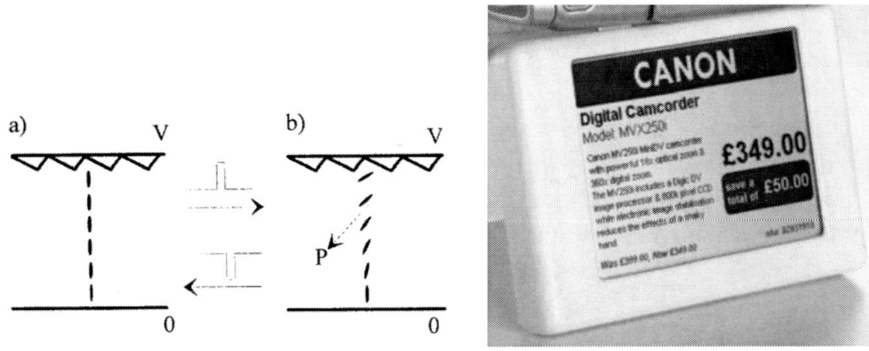

Figure 3.53. Bistable device made by special grating surface. (a)-homeotropic configuration; (b)-hybrid homeotropic to tilted alignment. Below typical ZBD device is shown [44].

The switching is made by favoring one of the states having a nonzero flexoelectric polarization with a voltage pulse of the same polarity [44]. (The opposite polarity pulse supports the homeotropic configuration (Figure 3.53). The grating surface was made by a photolithography technique and coated by a surfactant to induce a homeotropic LC orientation. There exist a possibility to optimize flexoelectric coefficients, anchoring energy and the grating profile in the bistable geometry as well as the optical characteristics of the device. The typical response times of the matrix addressed LCDs were found to be in a sub-millisecond range at U=10V [44].

The potential of the ZBD device has begun to be realised with its first commercial application. New technological advances are necessary, including understanding the relationship between grating shape and performance, the reproducible mass production of deep, sub-micron features on the inner surfaces of LCDs, and new liquid crystal material and addressing schemes that use the flexo-electric effect in nematic liquid crystals. The problem of ZBD device is rather high switching voltage and the problem of memorizing the intermediate levels, so only black/white bistable optical states become possible (Figure 3.53).

3.6.3. Optically Rewritable LCD

Optical rewritable technology (ORW) [45,46] is a modified method of azo-dye photoalignment that possesses traditional high azimuthal anchoring energy, up to $2{\cdot}10^{-4}$J/m^2, and has a unique feature of reversible in-plane aligning direction reorientation, i.e. rotation perpendicular to the polarization of an incident light. An ORW LC cell consists of two substrates with different aligning materials (Figure 3.54).

One aligning material is optically passive and keeps aligning direction on one substrate. The other aligning material is optically active and can change its alignment direction being exposed with polarized light through the substrate. In comparison with electrically controlled plastic display ORW can be significantly thinner and requires no ITO photolithography and etching on plastic substrate because no electrodes are needed. Switching and continuous grey scale are achieved by control of aligning direction of photoaligning azo-dye layer, which is insoluble in liquid crystal. By this mean one can obtain a specified twist angle in the ORW

LC cell that corresponds to the transmission level defined by the initial polarizers configuration (Figure 3.55). ORW is very tolerable to the cell gap variation as even 50% changing of the cell gap will not cause noticeable change in LC transmission value, while achromatic switching of all ORW grey levels can be obtained [45,46]. Every transmission level is stable and visualizes information with zero power consumption for a long time. Optical rewritable LC alignment is a good base for new types of LC displays, especially on plastic substrates, when no ITO films and electrodes are needed (Figure 3.55) [45,46].

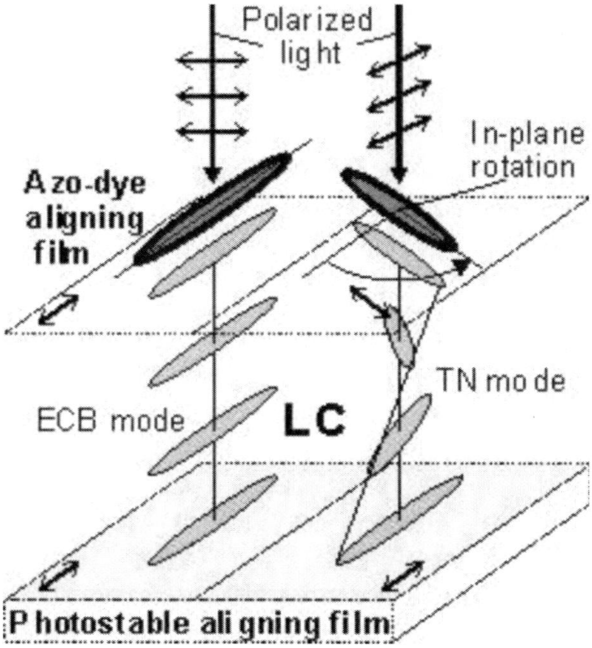

Figure 3.54. Operation principle of LC ORW cell [45,46].

No backlight is required as reflective type polarizer is used as the bottom substrate. The image is truly stable, can be written to any grey level with the saturation and rewritten a large number of times with a high reproducibility of properties. The overall reflection is about 30% (Figure 3.55), while the aspect ratio is 1. The unique property of being cell-gap non sensitive allows to maintain proper performance even when the device is bent. Fluorescent dye dopant of liquid crystal partly absorbs blue light and re-emit green light improving photopic reflection and enhancing color of the ORW E-paper (Figure 3.55). The ORW E-paper potential applications are light printable rewritable paper, labels and plastic card displays. The other ORW papers have been also recently developed, including (i) doping of photoaligning azo-dye to the bulk of cholesteric LC cell [47]; (ii) photosensitive chiral dopants in cholesteric LC capable to change the twisting power under UV exposure [20]; (iii) cholesteric LC layer with an organic photo-conductor layer [48]. However, these versions of ORW devices require sufficiently high electric field in combination with a writing light, which is not convenient.

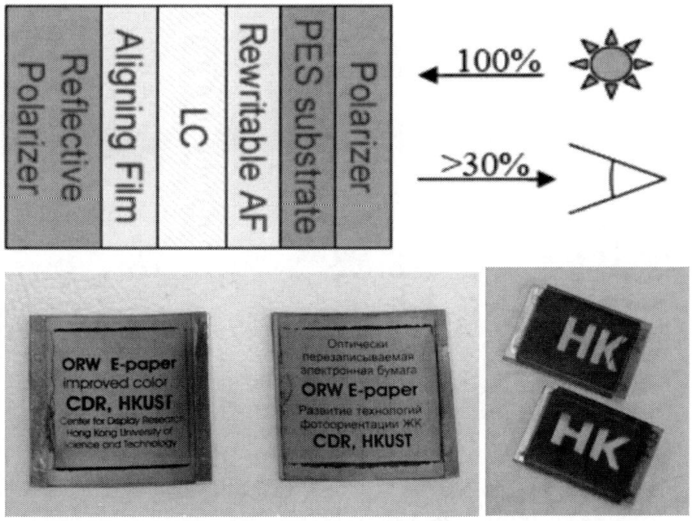

Figure 3.55. Optically rewritable E-paper. Top: basic structure; bottom: typical ORW image on flexible substrate [45,46]. ORW image based on LC mixtures with fluorescent dyes is also shown (bottom right corner).

3.7. LC Voltage Controllable Diffraction Gratings

Diffraction grating is one kind that has been well documented with either periodic amplitude or phase modulation. Among them, phase grating has the advantages of higher diffraction efficiency and better tunability, therefore most researches are focused on this type. We will consider the most promising of them in view of the switching speed and diffraction efficiency.

3.7.1. Nematic LC Diffraction Gratings

3.7.1.1. Liquid Crystal Gratings Based on Alternate Twisted Nematic (TN) and Homogeneous Planar (PA) Photoalignment

A design of tunable LC gratings with alternating 90° twisted nematic (TN) regions and homogeneous planar (PA) (45° with respect to the alignment direction of TN) is proposed, based on the combination of phase and amplitude modulation [49-51]. Both 1D and 2D diffraction gratings were demonstrated by periodic photoalignment of sulfonic azo-dye (SD1) [46] films with a linearly polarized light beam. The different voltage-dependent transmittances between adjacent regions make the device exhibiting four characteristic states. The combined modulation mode allows a very low manipulating voltage. Besides, the grating also has some advantages like simple and accurate fabrication, fast switching, high efficiency and high contrast ratio.

The photoalignment technique has been employed to accomplish the concept, and photoaligned azo-dye SD1 was utilized as the alignment layer [46,49-51]. The TN/PA grating cell preparation is schematically illustrated in Figure 3.56 [49-51].

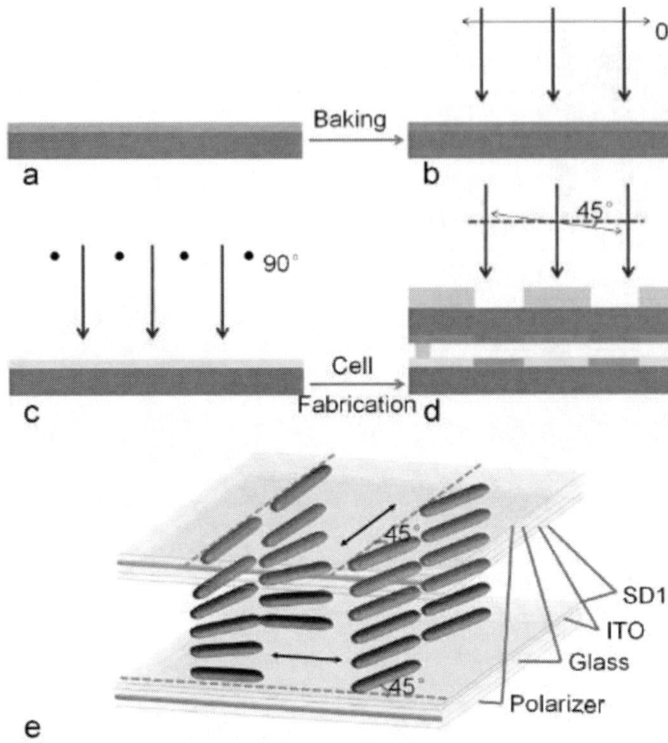

Figure 3.56. Schematic drawing of photoalignment procedure: (a) spincoating SD1, (b-c) orthogonal photoalignment on two cell substrates, (d) cell assembling and photopatterning with an amplitude mask, and (e) cell structure and LC orientations in alternate TN-PA domains [49-51].

The SD1 solution was spin-coated on ITO-coated glass substrates (Figure 3.56(a)). The two substrates were set orthogonal under linearly polarized light (Figure 3.56 (b) and (c)). Then the two substrates were assembled to form a 90°-TN cell. Afterwards, the cell was exposed again through an amplitude mask at 45° with respect to previous alignment direction. The exposed regions were realigned to PA, while the mask shadow domains still keep TN alignment (Figure 3.56(d)). Thus the cell with alternate PA and TN domains has been constructed. LC alignment in the cell is illustrated in Figure 3.56(e). The transmission-voltage (TV) curves were studied in the light path: a highly linearly polarized He–Ne laser (632 nm), which makes the system free of polarizer, the sample, an analyzer and a photodetector. The light beam was normally incident to the cell with its polarization direction parallel to front LC director in TN domains.

Total transmittance of the cell versus voltage reveals the contributions from both TN and PA regions respectively at different applied voltages, which is the overlap of TV curves of the two regions separately (Figure 3.57 (a)). The contrast ratio (voltage dependent intensity of +1st order divided by that of 0th) indicates the variations of diffraction efficiency along with external fields. The insets demonstrate the expected four states: stripe pattern (PA domains are dark), bright state, reverse pattern (TN domains are dark) and state approaching uniform dark respectively [49-51]. The transmission difference could still be observed in the bright state due to the limitation of the applying voltage step. The diffraction patterns could be observed at 5 V; further increasing the voltage to 15 V would induce a homogeneously dark state. The curve in Figure 3.57(b) exhibits a continuous intensity change of the first diffraction order. The extinction ratio (ER, intensity maximum over minimum) of the first order reaches over 120 in the whole voltage range (0-15 V). Furthermore, ER over 30 is achieved within the voltage interval of 0.15 V at very low operating voltage (0.80 V to 0.95 V), which permits a rapid switching [49-51]. The switching on and off time herein are defined as 10% to 90% in transmittance and reverse. The measured values of first order in the low voltage range were 6.5 ms and 36 ms separately. In addition to 1D grating, 2D diffraction gratings also fabricated, showing the expected 2D diffraction patterns [49-51].

3.7.1.2. Liquid Crystal Gratings Based on Orthogonal PA Photoalignment

LC gratings based on orthogonal PA alignment was also developed [50,51]. The adjacent orthogonal PA regions were still achieved by two-step photo exposure on photoaligned azo-dye SD1 [49-51]. The detailed fabrication process is shown in Figure 3.58. Just after coating SD1 layers, two glass substrates with ITO films were assembled to form a cell with 6.0 μm in thickness. The cell was exposed though linearly polarized light and rotated 90° and exposed again through an amplitude mask. The exposed window areas were realigned to be orthogonal to the original orientation of mask shadow regions. Thus the cell with alternate orthogonal PA alignments has been constructed. Based on above design, both 1D and 2D LC gratings were fabricated. Figure 3.58(b) and 3.58 (c) present the optical micro photographs of the cells under crossed polarizer.

A high total transmittance of 1D gratings up to 92% was achieved in the whole applying voltage range. Figure 3.59(a) reveals two voltage dependent transmittance (TV) curves of 0th and 1st orders respectively, which were calculated with respect to the total incident light intensity. As expected, the transmittance variations of 0th order and 1st order, as a function of applied voltage at fixed frequency of 1 kHz square wave, were mirror replica to each other [50,51]. The measured diffraction efficiency of 1st order was 31.4% at 1.75 V and optical contrast for 1st order was more than 150 in a voltage range from 0 to 15 V. The insets in Figure 3.59(a) provide some diffraction patterns at different voltages. The patterns follow the curves very well and directly prove the tunability of the gratings [50,51]. Figure 3.59(c) illustrates the voltage dependent behavior of the first order (n=1) for two input polarizations, one parallel to x axis, the other perpendicular. The result indicates that the test gratings have good polarization independence.

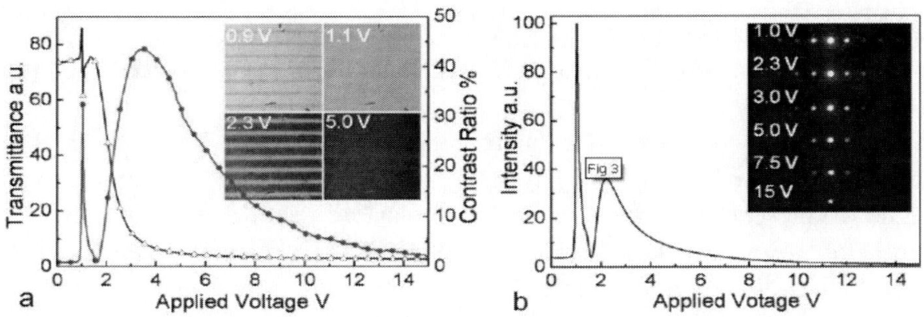

Figure 3.57. (a) TV curve of the cell (hollow triangles and line) and contrast ratio between voltage dependent intensity of 1st and 0th orders (solid circles and line). Insets show the four states respectively (all image sizes are 350×350 μm^2), (b) intensity of 1st order as a function of driving voltage. Insets show the diffraction patterns at different voltages [49-51].

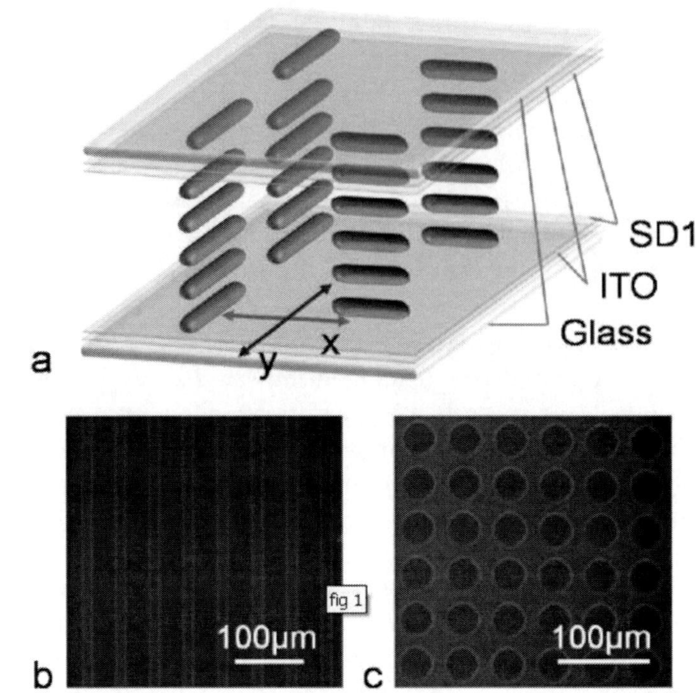

Figure 3.58. (a) Schematic cell structure of an orthogonal-PA cell, micrographs of (b) 1D and (c) 2D LC gratings [50,51].

3.7.1.3. Dual Frequency Addressed Hybrid Aligned Nematic(HAN) LC Grating

One of the disadvantages of nematic LC devices is the slow response time. To solve this problem, many approaches have been proposed. Among them, the dual frequency addressing has been proven to be an effective technique in some photonic devices. It would be beneficial if the the double frequence LC (DFLC), when the sign of the LC dielectric anisotropy is changed with frequency (Figure 2.9) could be used in LC gratings [51,52]. A patterned hybrid aligned nematic (HAN) cell was fabricated as shown in Figure 3.60(a). For the vertical aligned (VA) layer, a homeotropic polyimide was used; while for the patterned planar aligned (PA) one, photoaligned azo-dye SD1 was employed [46]. This substrate was exposed twice under a linearly polarized blue LED to form orthogonal alignments. The DFLC is infiltrated into the cell, with its physical properties listed as follows [51,52]: $n_{\perp} = 1.498$, $\Delta n = 0.191$ (at $\lambda = 633$ nm and T = 24 °C), $\Delta\varepsilon = 2.10$ at f = 1 kHz and –2.02 at f = 80 kHz, and $f_0 \approx 45$ kHz (see also (2.8)).

Figure 3.60 (c) ~ (e) shows the micrographs of the cell observed under a polarizing microscope. The experimental analysis was further proved through simulations [51,52]. Figure 3.60 (b) presents the cross-sectional (y-z plane) director distribution of the sample at voltage off state, from which the continuous change of LC orientations at the boundary was observed, as emphasized by two lines. Figure 3.60 gives (a) schematic illustration of the patterned HAN cell; (b) simulated LC director distribution in y-z plane with two black lines marking the boundary region; (c) ~ (e) micrographs of the sample, in each of which the relative directions of two polarizers compared with the stripes are labeled in arrows.

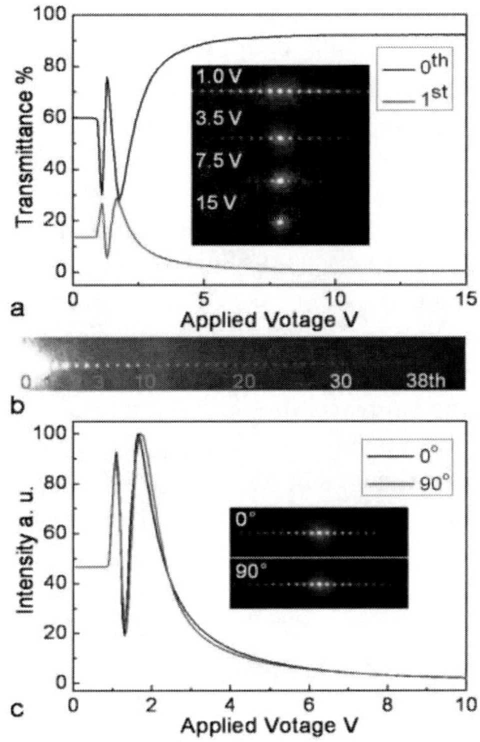

Figure 3.59. (a) TV curves of 0th and 1st orders, insets show the diffraction patterns at different voltages, (b) diffraction patterns of 1st order at voltage off state, (c) diffraction intensities of 1st order under different polarized incident light, insets show corresponding diffraction patterns at voltage off state [50,51].

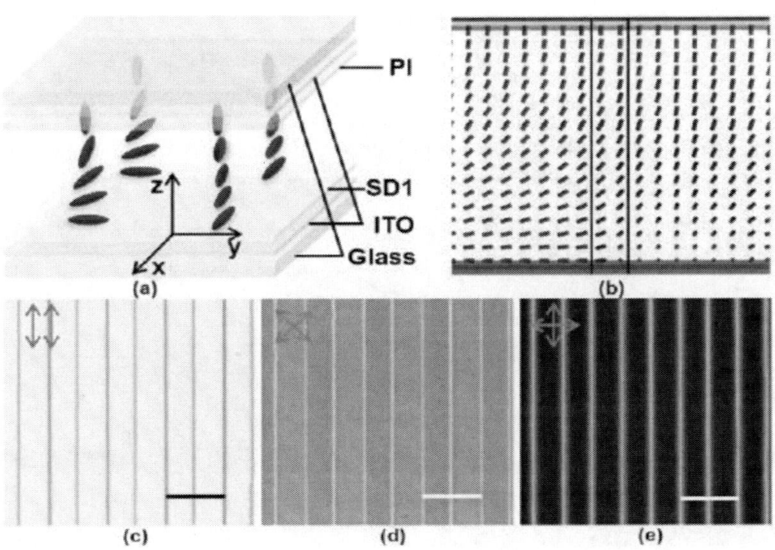

Figure 3.60. (a) schematic illustration of the patterned HAN cell; (b) simulated LC director distribution in y-z plane with two black lines marking the boundary region; (c) ~ (e) micrographs of the sample, in each of which the relative directions of two polarizers compared with the stripes are labeled in arrows [51,52].

Figure 3.61 exhibits two voltage dependent intensity curves of 1st order measured when rectangular electric signals are applied at f = 1 kHz and f = 80 kHz respectively. The sample was placed in three different directions: 0°, 45° and 90° relative to the incident polarization. The curves measured at the same frequency are in a high accordance with each other, showing good polarization independence. As respected, at saturation voltage, the intensity was suppressed at low frequency while keeping strong at high frequency. Thus the switching effect could be achieved through frequency alternating at a given voltage above saturation. The extinction ratio exceeds 20 dB, which is good for some applications [51,52]. The insets visually reveal the high switching performance. Further improvement on extinction ratio is achievable by more precise cell gap control and employing a quarter-wave plate that can greatly eliminate the residual phase thus reducing the intensity at off state.

Submillisecond switching has been achieved, with measured switch on/off time of 350 μs and 600 μs respectively, where on/off time was defined as the duration time that intensity changes from minimum to maximum and reverse (Figure 3.62). Further improvement could be accomplished by increasing voltages. For instance, the on/off time reduced to 280 μs and 460 μs respectively at the higher applied voltage [51,52].

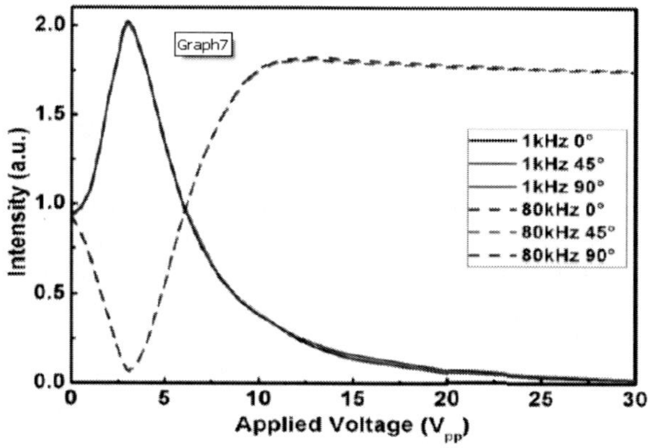

Figure 3.61. Voltage dependent intensities of 1st order at 1 kHz and 80 kHz when the sample placing at three different directions: 0°, 45° and 90° relative to the incident polarization [51,52].

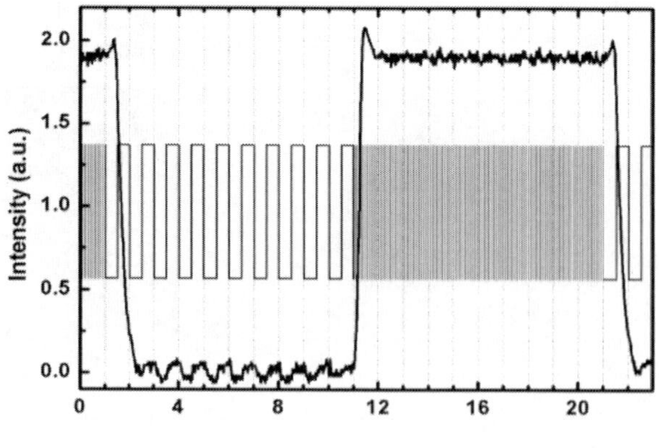

Figure 3.62. Switching performance of the dual frequency addressed HAN LC grating sample [51,52].

3.7.1.4. Switchable Liquid Crystal Grating with Sub Millisecond Response

Liquid crystals (LC), because of their easy tunability by electric and magnetic fields, became one of the very important elements in various photonic devices. Switchable grating is one of these elements which have been studied broadly. To increase the speed of the nematic LC devices, several methods, viz., fringe field effect, guest host effect, etc, have been proposed so far but the response time is typically limited to 1ms [1-3]. Recently, a fast switchable grating based on polymer stabilized short pitch chiral LC was proposed with the switching time for this grating is 600 μs, whereas the required electric field is around 20 V/μm, which is relatively high [53]. Another approach to speed up such devices has been proposed by utilizing the flexoelectric effect of short pitch cholesteric LC [54]. The response time is limited to 200 μs; however, this technology has several material issues and a very complicated fabrication procedure.

A relatively simple and faster LC device that manifests saturated electrooptical (EO) modulation up to very high frequency (~2 kHz) and, thus, could find application in several fast photonic and display devices was also proposed [55]. A switchable LC grating was based on the fine striped three-electrode structure and a special driving scheme to reduce the LC relaxation time (3.11). The simultaneous application of vertical and horizontal electric fields on planar aligned nematic LC was applied to reduce both on and off LC switching time. For this purpose, a very fine ITO pattern has been designed and the cell structure for one period of the cell is shown in Figure 3.63 [55]. On one substrate (bottom), the ITO was patterned similar to in plane switching (IPS) substrate with the electrode width and gap between two electrodes of 2 μm, whereas continuous ITO has been used for the other substrate.

In order to generate the switchable grating profile in the cell, both vertical and in-plane electric fields have been applied [55]. The simulated LC director profile in the dark and bright state is presented in the Figures 3.63(b) and 3.63(c). When in-plane electric field is applied, the LC molecules between the in-plane electrodes switch very fast to the planar orientation. However, the region just above electrodes remains vertical, which creates the periodic variation of the refractive index and, thus, generates the diffraction grating profile. Figure 3.64 represents the intensity profile of the diffraction pattern at the electric field of 10 V. When the in-plane electric field has been applied, the clear diffraction pattern appears (Figure 3.64(a)), whereas no diffraction has been observed (Figure 3.64(b)) on the application of the vertical field. To elaborate the phenomenon, the surface intensity profile of the cell has been plotted in the Figure 3.64(c) that reveals the grating profile for the bright state, while for the black state, it is perfectly flat. Thus, the diffraction profile can be switched by applying alternative vertical and in-plane electric fields. The diffraction profiles with the in-plane and vertical fields have been shown in Figures 3.64(d) and 3.64(e), respectively. The optical contrast ratio for the first order diffracted beam, at the electric field of 10V, was 400:1, whereas the total diffraction efficiency was 21% [55]. The contrast ratio and the diffraction efficiency can further be improved at higher electric field and by optimizing the LC and cell parameters.

Figure 3.65 reveals the response time dependencies on the applied voltages at a frequency of 500 Hz. Both the response times decrease at the higher electric fields. The on and off times at the electric field of 10 V/lm were 95.6 μs and 151 μs, respectively. This behavior can be explained on the basis of Eq. (3.10). The solid lines in the Figure 3.65 represent the best theoretical fits of the Eq. (3.10). The response time is around 150 μs that gives us an opportunity to drive the proposed grating at very high frequency. The inset of the Figure 3.65 shows the EO response for the first order of the diffracted beam at 1.5 kHz, which manifests the saturated optical states even at higher frequencies.

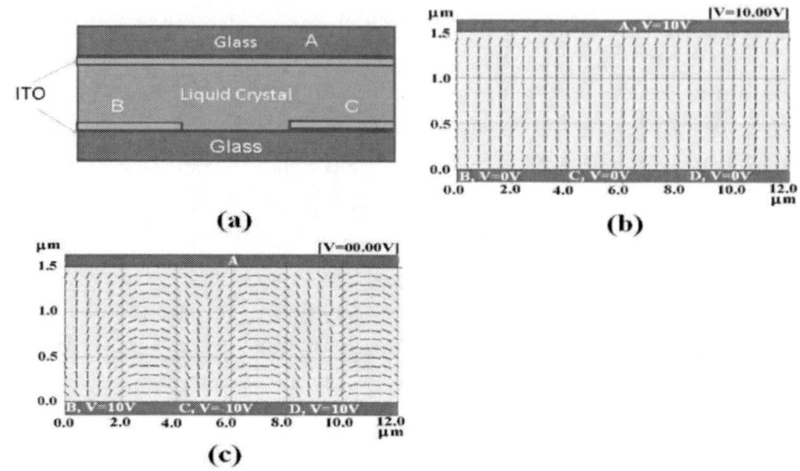

Figure 3.63. (a) Detail description of one period of the used three electrode LC cell. A, B, and C are the electrodes where the electrode width and gap for B and C is 2 µm. (b) represents the simulated LC director profile of the dark state where 10V of electric field is applied on the top electrode, i.e., A. (c) The LC director profile when alternate electric field has been applied between B and C [55].

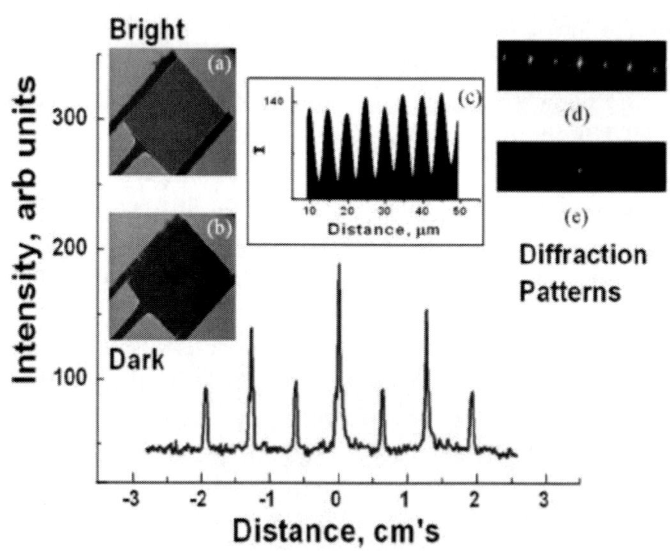

Figure 3.64. The intensity profile along the diffraction pattern at electric field of 10 V. (a) and (b) represent optical photographs of the cell under crossed polarizer in bright and dark states, respectively. (c) shows the surface intensity profile of the cell in the bright state, while (d) and (e) show the diffraction pattern in bright and dark states, respectively [55].

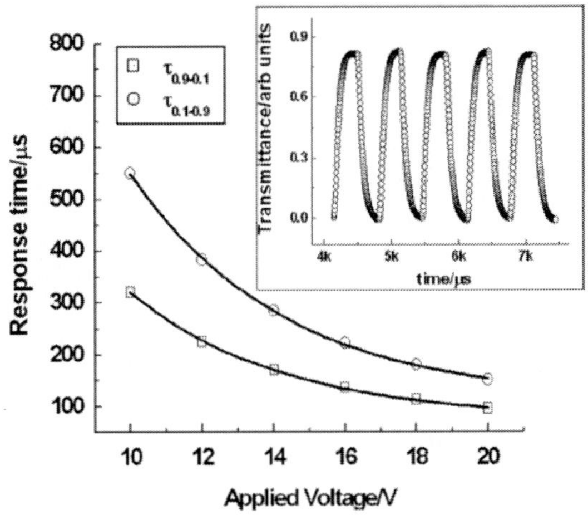

Figure 3.65. The applied voltage dependency of the response time at the fixed frequency of 500 Hz. Inset shows the EO response of the grating cell at the applied voltage of magnitude 20V and frequency 1.5 kHz [55].

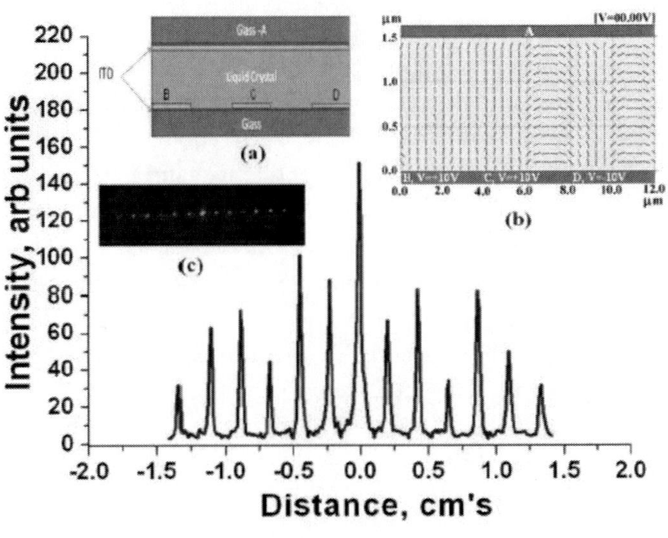

Figure 3.66. The diffraction intensity profile, when the electric field is applied between electrodes C and D. (a) The cell structure with three bottom electrodes (i.e., B, C, and D), (b) shows the director profile of the LC, and (c) the optical photographs of the diffraction pattern [55].

This fast EO response makes this technology to be applicable for wide range of tunable photonic and other EO elements, viz., tunable grating, optical interconnects, beam steering devices, optically addressed spatial light modulator (SLM), etc. Another advantage of this method is the switch-ability of grating periods. By increasing the number of electrodes on the IPS patterned electrode, one can have several grating periods. An example with three IPS electrodes has been shown in the Figure 3.66(a). The first two consecutive electrodes have been kept at the same potential, while it is different for the third electrode. The director profile of bright (diffractive) state under this condition has been shown in Figure 3.66(b), which reveals the grating period of 16 μm. On the other hand, the director profile for the black state is the same as discussed earlier. Figure 3.66(c) shows the diffraction intensity profile for the grating with 16 μm. Therefore, increment in the number of electrodes may offer higher number of grating periods.

Further optimization in the LC and cell parameters can improve the diffraction profile. Thus, this technology can find application in high-resolution 2D or 3D displays, optical shutters and photonic devices, and very promising for applications, demanding sub-millisecond timescales.

3.7.2. Other LC Switchable Diffraction Gratings

3.7.2.1. Ferroelectric LC (FLC) Diffraction Grating

FLC diffractive elements have an edge on all of the existing technology because of their fast switching speed and low power consumption [1-3]. FLC gratings was fabricated characterized by a high frequency saturated electro-optical (EO) modulation, low driving voltage, high diffraction efficiency, and high contrast ratio [56]. Such gratings are very promising for applications in a variety of fast photonic and display devices. Electrically suppressed helix (ESH) mode was used for FLC diffraction grating [38,56].

The patterned photoalignment alignment method [46] was used to generate the periodic distribution of refractive index and thus the switchable grating with contrast more than 7000:1 for the first diffraction order and diffraction efficiency about 68% [56]. Since the response time was faster than 20 μs it could provide optically saturated EO modulation at high frequency up to 5 kHz [56]. FLC grating cell has been prepared to meet this condition. The FLC FD4004N (From Dai-Nippon Ink and Chemicals (DIC)) having spontaneous polarization Ps = 61 nC/cm^2 and tilt angle $\theta = 22.05^0$ was chosen and sulfonic dye SD1 (DIC) [46] was used for the photoalignment. In order to provide the periodic distribution of refractive indexes, the FLC has been aligned in two orthogonal planar alignment regions by mean of two step UV exposures. The schematic diagram of the cell structure that illustrates two orthogonal alignment regions and the optical microphotographs is shown in Figure 3.67. Figure 3.67(a) illustrates two different overlapped alignment regions having mutually perpendicular alignment directions. Figure 3.67(b) shows the optics of the two alignment domains in between two crossed polarizer with one switching position parallel to either polarizer or analyzer. It can be seen that no diffraction occurs in the dark state when the switching position of the two alignment domain is either parallel to the polarizer or analyzer. However, the diffraction appears for the other switching position where the optic axis for these two domains makes an angle 2θ and 90-2θ for the polarizer, where $\theta=22.05^0$ is the FLC tilt angle. Figures 3.67(c) and 3.67(d) represents the optical microphotographs of the FLC grating structure for 1D and 2D patterns made through the crossed polarizers at the electric field E=6V/μm. The period of both gratings P_g was 50 μm, which, if necessary, can be reduced down to 1 μm provided that the FLC helix pitch is sufficiently small.

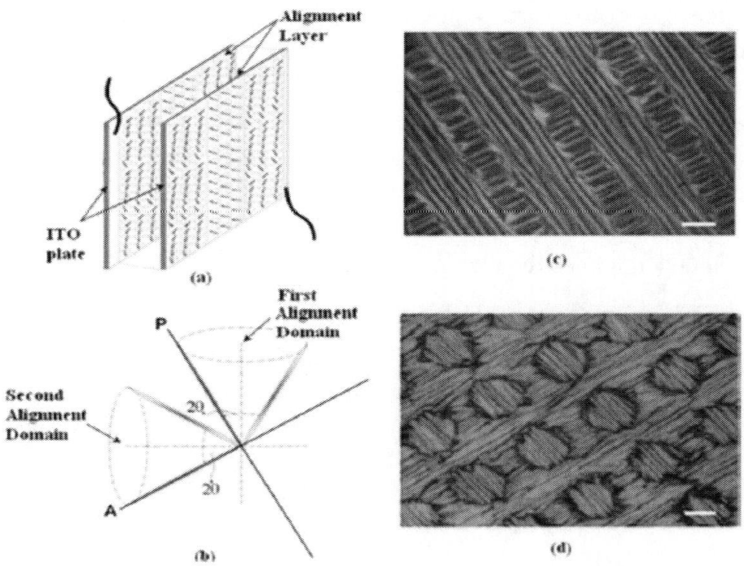

Figure 3.67. (a) FLC grating cell with two orthogonally aligned domains where the smectic layer normals arranged to be parallel to the substrate are perpendicular to each other; (b) Orientational structure in the adjacent domains and the geometry of crossed polarizers (P and A stand for the polarizer and the analyzer, respectively); (c)–(d) Optical microphotographs of 1D and 2D FLC gratings taken at V=10 V. The length of the white marker is 15 μm [56].

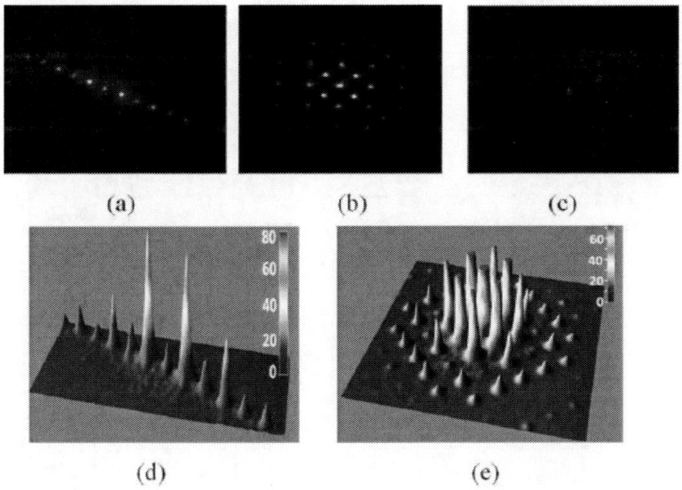

Figure 3.68. Diffraction profiles of FLC grating: (a) the diffractive state of 1D grating; (b) the diffractive state of 2D grating; (c) the black state, which is same for both gratings. (d) and (e) Intensity profile measured at the voltage V=5V for 1D and 2D gratings, respectively [56].

The diffraction profile for 1D and 2D grating is shown in Figures. 3.68(a) and 3.68(b), respectively [56]. Figure 3.68(c) shows the dark state of the FLC grating, which is the same for both 1D and 2D structures. This figure demonstrates that the good optical quality is combined with the high contrast between diffractive and non-diffractive states. The corresponding intensity profiles are plotted in Figures 3.68(d) and 3.68(e) for 1D and 2D grating, respectively [56].

Good diffraction efficiency of FLC grating was also demonstrated [56]. Figure 3.69 presents the results for the transmittance of the first diffraction order, which is plotted as a function of voltage in both the bright and dark states. It can be seen that both curves reach the regime of saturation at voltages higher than 2 V. Similar voltage dependence was also observed for the transmittance of the zero order (see insert in Figure 3.69). The diffraction efficiency is thus independent of the electric field higher than 1.3V/μm.

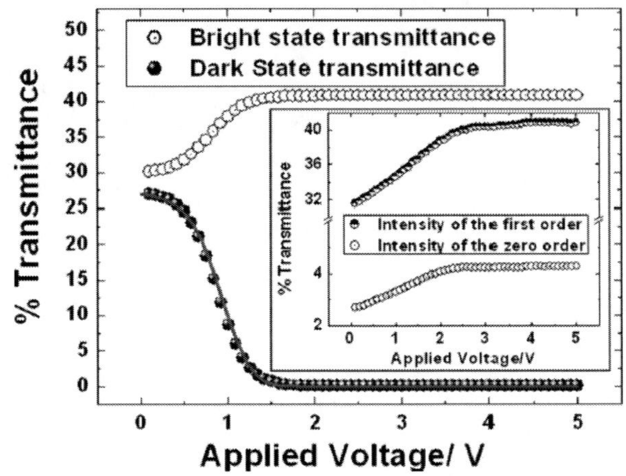

Figure 3.69. Electric field dependence of the transmittance for the first diffraction order. Insert shows the bright state transmittance of the first and zero diffraction orders as a function of voltage. Solid line represents the theoretical values [56].

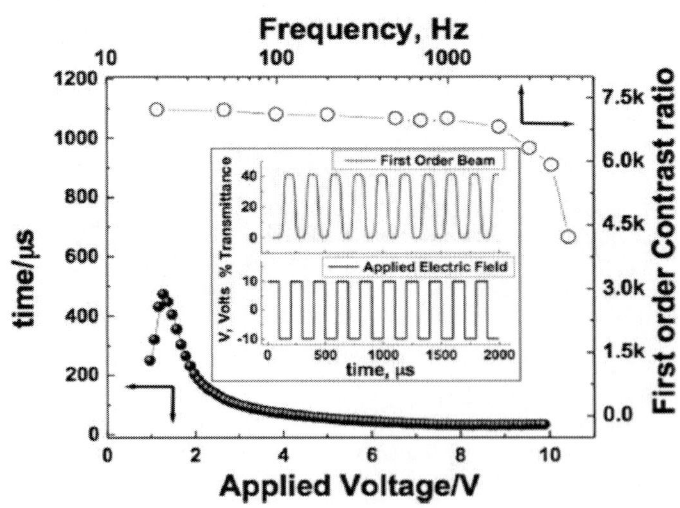

Figure 3.70. Voltage dependence of the response time at driving frequency of 500 Hz (solid circles) and frequency dependence of the first order contrast ratio at the fixed applied voltage V=10V (open circles). Insert shows time dependence of the applied voltage (bottom) and the EO response of the first diffraction order (top) at the wavelength λ=632nm and at the operational frequency f=5 kHz [56].

The results on the dynamics of the FLC grating are shown in Figure 3.70 [56]. The electric field dependence of the response time of the first diffraction order is similar to the one for the planar aligned FLC cell. It reaches the maximum value at the critical field of FLC helix unwinding (3.30) and decreases when the electric field further increases.

The insert in Figure 3.70 presents the EO response of the first diffraction order measured at frequency of 5 kHz. It is clearly characterized by the optically saturate bright and dark states with high contrast. Since the contrast ratio for FLC cell with ESH mode is typically very high, similarly high contrast has been observed for the first diffraction order. The contrast ratio for the first order (i.e., the ratio of the I_{max}/I_{min} for the first order) is more than 7000:1 at the electric field of the 6V/μm [56]. The contrast ratio does not show a pronounced decrease up to the frequency of 2 kHz. Even though it is considerably reduced at higher frequencies, the contrast ratio for the first diffraction order is 4200:1 at f=5 kHz.The proposed grating have extreme potential for application in many devices, e.g., projection displays, micro-displays, beam steering devices, and other photonic and adaptive optic elements like optical interconnect.

3.7.2.2. Polymer Dispersed LC (PDLC) Diffraction Grating

The holographic technique may be used to form periodic diffraction gratings in PDLC films [57,58]. Figure 3.71 shows the principal set-up to make a holographic recording in PDLC films. A mixture of a light-curable pre-polymer and nematic LC are placed between two ITO coated glass plates and two laser beams are irradiated onto the sample. The interference fringes of the laser beams provide the LC/polymer periodic structure by photo-polymerization induced phase separation.

In the off state the periodic modulation of the PDLC refractive index in holographic PDLC (HPDLC) films results in a resonance reflection of a specific wavelength from the illumination according to the Bragg condition, which is given by (3.28) for a normal light incidence (Figure 3.71,a). When an electric field is applied, the refractive index of LC droplet matches that of the polymer, HPDLC becomes transparent and no reflection occurs (Figure 3.71,c). The HPDLC cell can thus change the reflection at specific wavelengths without loss of the transmission at other wavelengths. A single LC/pre-polymer mixture and a single laser (e.g. Ar^+ laser,488 nm) can be used to fabricate the three basic color HPDLC films by changing only optical setup, including the angles of the two laser beams incidence[57] (Figure 3.71,b). Stacking the three HPDLC cells together a color reflective HPDLC device can be obtained with the following characteristics [57-59]:

(i) high reflection efficiency (15-25% in a white and 70-80% in a monochromatic light);
(ii) considerably large contrast ratio (20-30:1), however, at high driving voltages (about 40V at 10μm cell) ;
(iii) high color purity (10nm spectral width on the three basic wavelength); sub – millisecond switching.

Recently, most researchers have focused on the fabrication of two-dimensional (2D) electrically-tuneable diffraction grating on PDLC film [60-62]. The 2D grating can diffract incident light in various directions. It can be widely applied in optical beam steering, beam splitters or phase modulation devices [60-62].

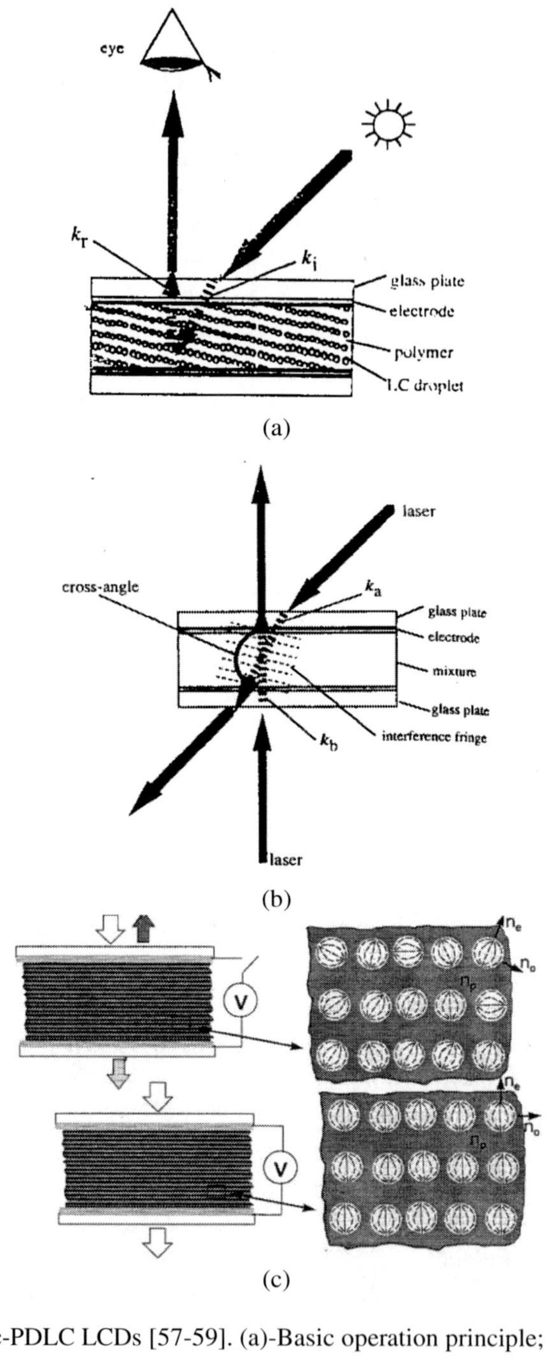

Figure 3.71. Holographic-PDLC LCDs [57-59]. (a)-Basic operation principle; (b)- Fabrication set-up for color reflective HPDLC; (c)- Switching in electric field [57-59].

3.8. VARIOUS LCD ADDRESSING SCHEMES

LCD image can be tunable by altering the liquid crystal orientation in certain regions, creating an optical contrast through the anisotropy in the liquid crystal's refractive index and/or optical absorption. The electrical addressing of LCD with a high information content can be subdivided into: (i) active matrix addressing; (ii) passive matrix (PM) addressing.

In most active matrix LCDs a discrete nonlinear device, such as a diode or transistor, is associated with each pixel of the display, which are regularly arranged in a rectangular array or matrix. Figure 3.72 shows a diagram of the most common active matrix LCD, the thin film transistor (TFT) display.

There are three types of electrodes present on this display: control electrodes, signal electrodes and pixel electrodes. For TFTs the control electrodes operate the gates of the transistors to momentarily switch the transistors into their conducting states so that the voltage present on the signal electrodes can be channeled to the appropriate pixel electrodes at the proper times. The pixel is defined by the region of overlap between the pixel electrode and a single, common counter electrode positioned on the opposite side of the liquid crystal layer.

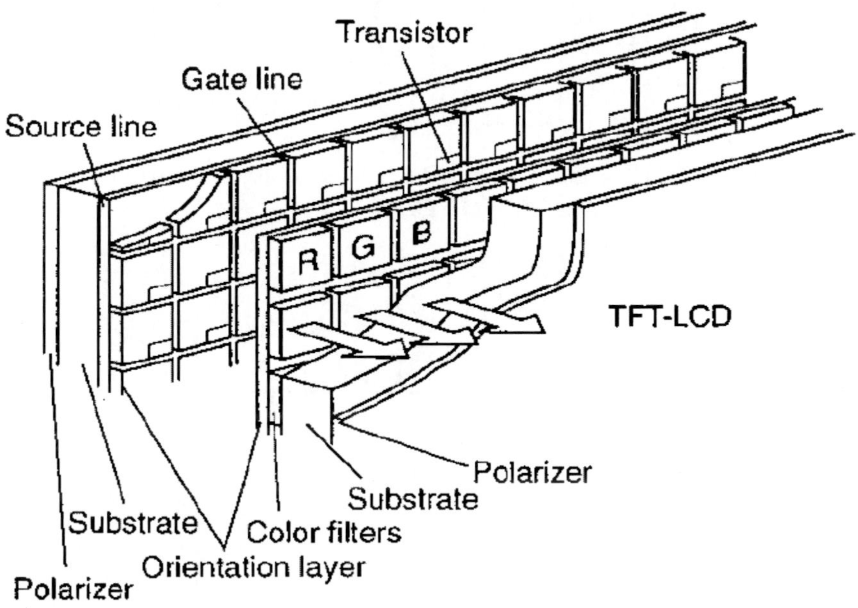

Figure 3.72. The structure of active-matrix display.

In a passive matrix display there are no control electrodes since there are no nonlinear devices to control; and the signal electrodes are the same as the pixel electrodes. Every pixel is uniquely determined by the region of overlap of a row and column electrode, making it possible to access N·M pixels with only N+M electrodes. For a passive matrix LC device it follows that the voltage across the liquid crystal, which determines the optical state of the pixel, is the potential difference between the voltages applied to the corresponding row and column electrodes (Figure 3.73). We have provided Figure 3.73 below, which indicates the principles of passive matrix LCD.

It is important to realize that a drive waveform applied to a column electrode influences the pixel waveforms appearing across every pixel in that column.

In order to display the desired pattern of pixels it is therefore necessary to ascertain the characteristics of the pixel waveform that the electro-optical effect responds to so that this can be exploited by the addressing technique. In the surface stabilized ferroelectric liquid crystal display (Figure 3.35), for example, the polarity of the last applied pulse whose amplitude-time product lies above a certain threshold value determines the optical state of the pixel (Figure 3.74).

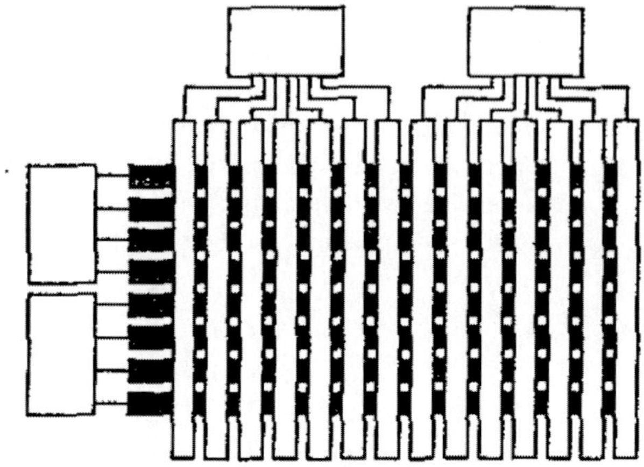

Figure 3.73. Common passively addressed matrix LCD. The row drivers (left) and column drivers (above) are used to address every pixel of LCD matrix.

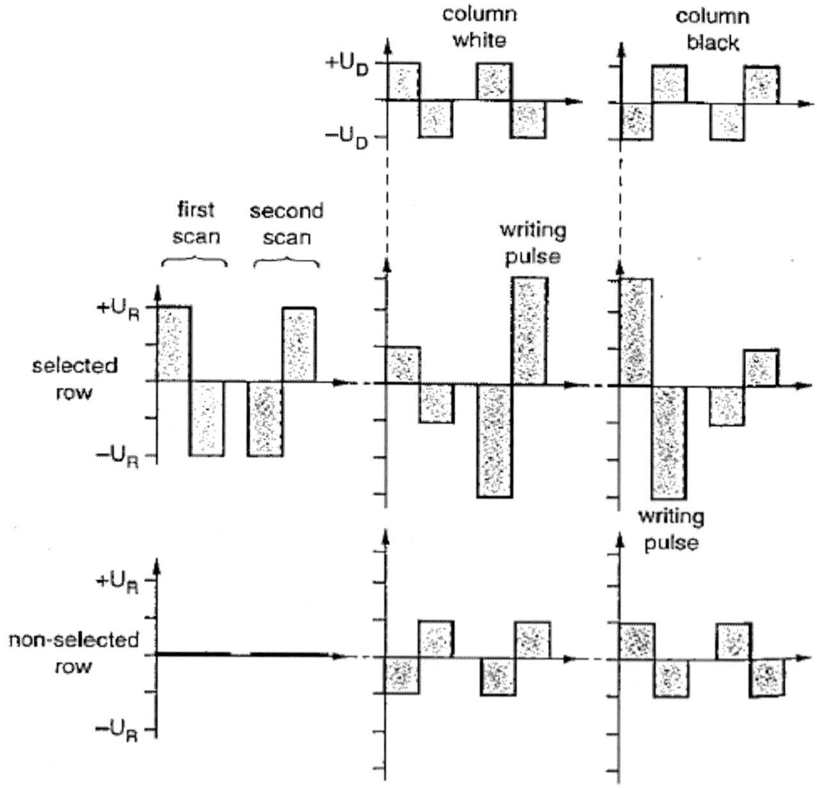

Figure 3.74. Seiko multiplexing scheme for passive addressing of FLC displays [1].

Other electro-optic effects such as the TN, STN and LCD effects respond, over a wide range of frequencies, only to the root-mean-square (rms) vaiue of the pixel waveform (Figure 3.75).

Some new perspective methods of LCD addressing is color sequential LCD. The principle of color sequential LCD is shown in Figure 3.76. The driving of such color sequential LCD includes three sub-frames. The response time of the LCD needs to be very fast in order to fit the field sequential driving method. Suppose that the driving frequency is 60 frames per second, i.e. the frame duration is 16.67 ms. Each sub-frame will have 5.5ms. Again, each sub-frame has to further divide into three parts. The first one is the data loading time which is about 0.5ms. Secondly, it comes to the LC response time which is about 2.1 ms. Finally, the total LED illumination time is around 2.9 ms (Figure 3.76).

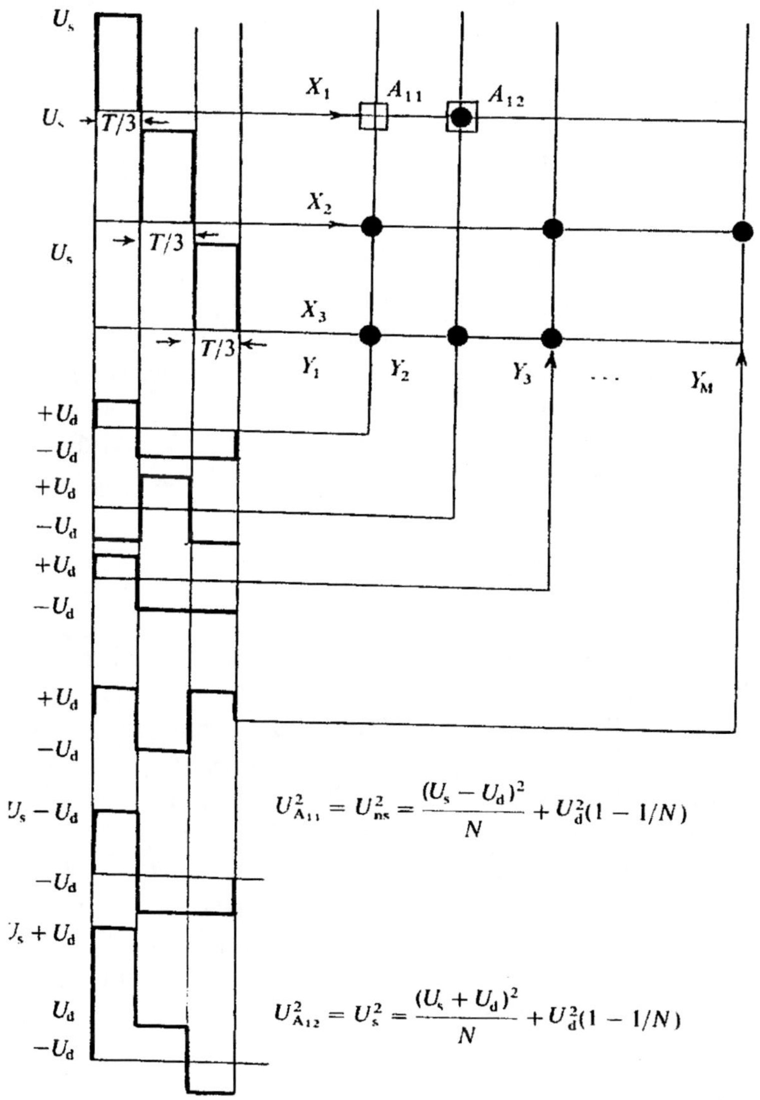

Figure 3.75. Principles of matrix addressing of LCD with N rows (N=3) and M columns [1].

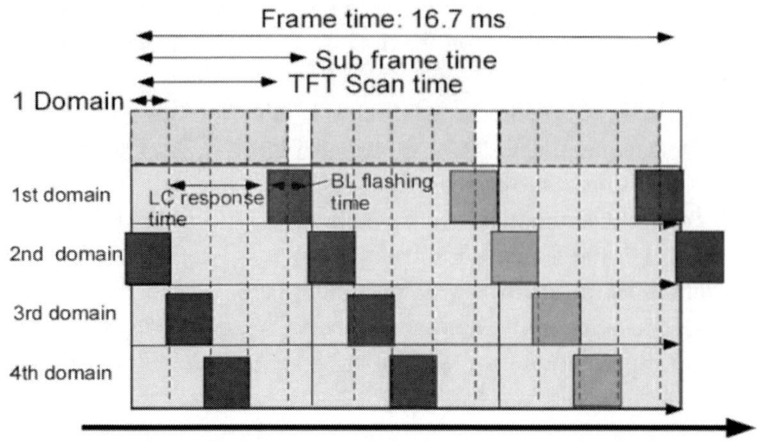

Figure 3.76. Principle of LCD color sequential switching.

Let us consider the LCD quality factors as a function of LC mixture parameters and its anchoring properties for various types of LCD described above with different addressing.

3.8.1. Passive Matrix Addressing

One of the basic limitations of passive matrix LCD (Figure 3.75) is a so called "iron law", which states that the resolution of the LCD strongly correlates with the steepness of the transmission-voltage / reflection-voltage (TVC/RVC) curves. As mentioned above, the steepness of the iron law considerably restricts the quality of passive matrix LCD (PM-LCD), especially for a sufficiently high resolution (Figure 3.77). The selection ratio, shown in Figure 3.77

$$V_{on}/V_{off} = 1+p, \tag{3.77}$$

where p- is a steepness parameter, V_{on} and V_{off} corresponds to the switched on (switched off) state of PM-LCD respectively.

Both elastic, dielectric and optical parameters of LC mixtures, as well as LC cell configuration (pretilt angle, twist angle, anchoring energy) are very important [1-3]. All the properties of highly multiplexed PM TN and STN-LCD such as contrast ratio, brightness, response time, temperature range etc. are getting worse with increasing the resolution (number of addressing lines), as to keep a high steepness of TVC curve is a challenging task.

The LCD modeling software [4,7-9,36] provides an efficient tool for modeling and optimization of PM-LCD electrooptical performance based on TN, STN and other electrooptical modes in LC mixtures dependent on the LC mixture parameters and cell configuration. Some results of PM-LCD optimization based both on calculations and experiment for TN and STN effects are shown below.

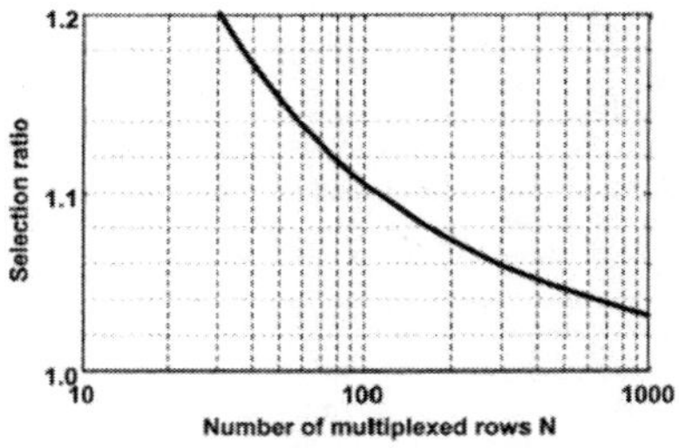

Display resolution	Selection ratio, single	Selection ratio, dual scan
160x160	1.0824	1.1188
VGA (640x480)	1.0467	1.0667
SVGA (800x600)	1.0417	1.0595
XGA (1024x768)	1.0368	1.0524
SXGA (1280x1024)	1.0318	1.0452
UXGA (1600x1200)	1.0293	1.0417

Figure 3.77. Iron law of passive matrix addressing LCD: the number of the addressing lines strongly correlates with the selection ratio, $V_{on}/V_{off,}$ where V_{on} and V_{off} corresponds to the switched on (switched off) state of PM-LCD respectively.

1. A proper choice of "back flow" viscosities considerably shortens PM STN-LCD response time (Figure 3.78).
2. Viewing angles of PM TN-LCD are better in a normally black mode (Figure 3.79).

The interest to PM LCD decreases recently, as all the high resolution displays for most applications, such as TV, desk-top and PC monitors, mobile phones, PDAs are active matrix LCD (AM-LCD).

3.8.2. Active Matrix Addressing

Thin film transistors (TFT-LCDs) were considered to be the basic structure for active matrix addressed LCD (AM-LCDs) (Figure 3.72). TFT operates as a very good conductor when the gate (row) voltage is large and a very poor conductor in the case of small gate voltages [1-3] (Figure 3.80). When the row voltage is zero, no matter how high the column voltage is, the current will not flow through the LC pixel element. Opening the gate voltage by a selected row pulse makes it possible or deliver to the LC pixel element a definite space charge, proportional to the required contrast.

The LC pixel is connected to the ground and, consequently, the space charge is allowed to relax during the frame interval, with the characteristic time defined by the capacity C_{LC} and resistance R_{LC} of the LC layer (Figure 3.80). A zero DC voltage in LC element is provided by a corresponding sign alternating of the column voltages during the subsequent frame times. The operation is allowed due to the symmetry of the TFT conductance in both directions. There is no need to change polariry of the row-addressing voltages, as their only function is to open the gate (Figure 3.80). A small capacitor Cs is added to the LC pixel capacity C_{LC} to retain the contrast during the frame time.

TFTs are certainly advantageous since (i) the crosstalk is practically absent, because the column voltage hardly affects LC pixels during the non-addressed period, and (ii) grayscale generation is much easier and resembles that of directly addressed LCD elements. However, the aperture ratio of TFT AM-LCDs (i.e., the ratio of the dead area on pixel), where TFT is located to the total active area is lower than the total area of the pixel.

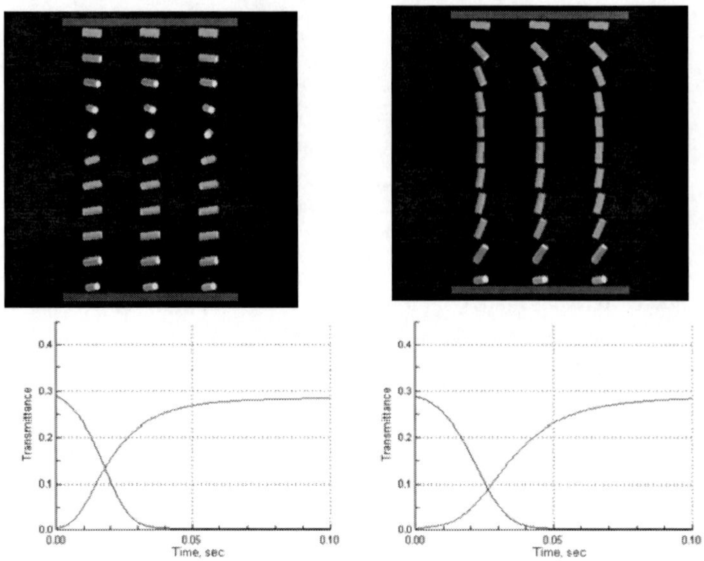

Figure 3.78. Simulation of PM STN-LCD. 20% decrease of the response time for a proper choice of a shear (back flow) LC viscosities (see 2.28). Top: off (left) and on (right) states of STN-LC cell. Bottom: STN LCD response: optimized (left), and common (right).

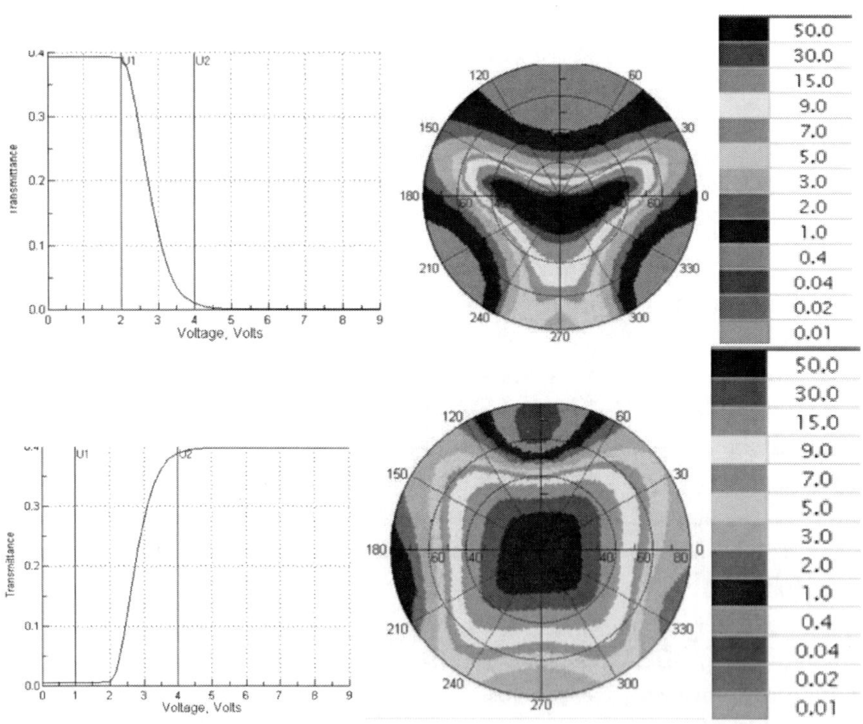

Figure 3.79. Viewing angles of normally white (above) and normally black (below) TN-LCD. The TVC curves of the modes are shown in the left. The viewing angles are defined by contrast ratios at different viewing angles.

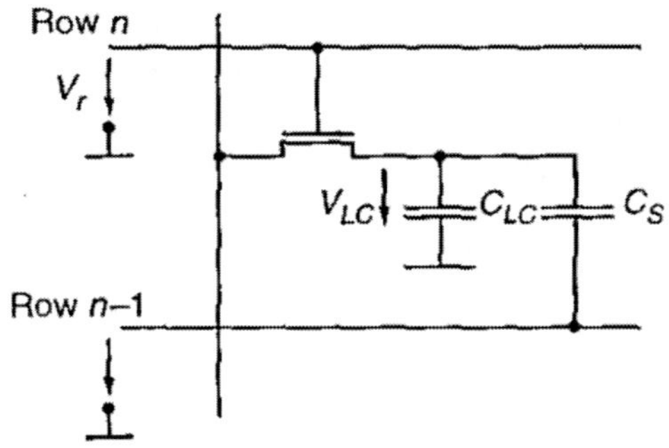

Figure 3.80. Equivalent electrical circuit of TFT AM-LCD pixel [1-3].

The requirements for a TFT are derived from the number *N* of rows and the frame time T in which a picture is written in. The row is addressing during the interval $\tau = T/N$. In order to guarantee a proper charging of the capacitors $C_{LC} + C_S$ during the frame time T, the following conditions for the time intervals must take place [1-3]:

$$R_{on}(C_{LC} + C_S) < 0.2\tau = 0.2\ T/N \tag{3.78}$$

where the factor 0.2 means 1% error in charging voltage [1].

The stored charge has to be held during the frame time T to avoid flickering of the image. This enables us to write (1 % error in holding voltage):

$$R_{off}(C_{LC}+C_S) \geq 100T \tag{3.79}$$

Taking into account (3.78) and (3.79), we have for the TFT resistance ratio:

$$R_{off}/R_{on} \geq 500N \tag{3.80}$$

If Ioff ~ 10^{-12} A and N = 1000 rows, we must have $I_{on} = I_{off}R_{off}/R_{on} \geq 0.5\ \mu A$.

One of the important characteristics of the TFT-LCD is a voltage holding ratio (VHR). VHR is defined as the voltage at TFT-LCD pixel at the end of the frame period, divided by the voltage initially set to it during the row addressing time (Figure 3.81).

$$\text{VHR}\ (\%) = \frac{\int_0^{16.67ms} V(t)dt}{5V \cdot 16.67ms} \cdot 100\% \tag{3.81}$$

VHR strongly depends on the decay time of the voltage on the pixel element to the frame time T. If VHR is too small, not only will the pixel voltage become smaller it will also cause flickering. Thus the brightness of the pixel changes periodically. This is highly undesirable.

The voltage holding ratio (VHR) is a critical electrical parameter for LCDs. It is a measure of the amount of ions in the bulk of LC. In TFT-LCD configuration VHR shows how fast the charge will relax on LC pixel, while the other LCD rows are being addressed. Too fast relaxation results in the flicker effect. The VHR value can be measured as follows (Figure 3.81). Firstly a voltage pulse of 5V was applied the LC cell during 64 μs (260 lines in 16.67 ms frame time, Figure 3.81); and then the voltage V(t) dropped down during a period of T = 16.67 ms (frame frequency about 60 Hz) after the pulse was switched off at t = 0 (Figure 3.81). The acceptable value of VHR for TFT-LD applications is more than 99% for both room temperature and 80^{0}C.

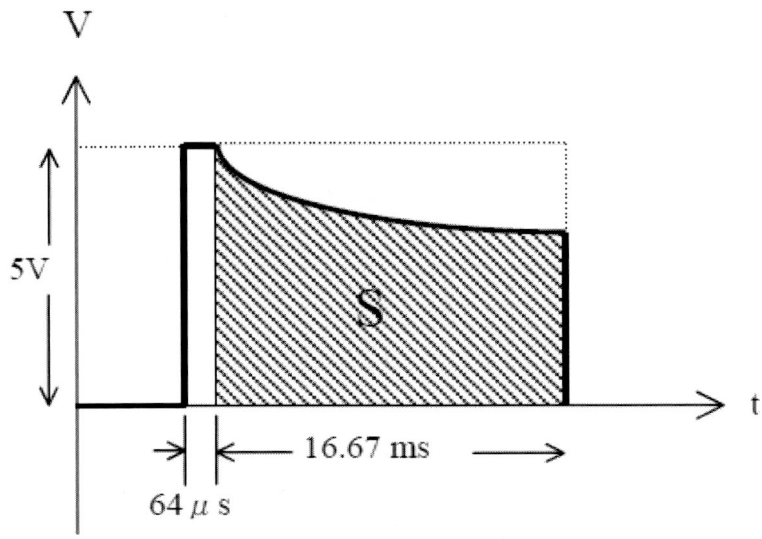

Figure 3.81. Measurements of VHR: voltage pulse of 5V is applied during 64 μs to LC cell. Using this addressing time 260 lines of AM-LCD can be addressed during the frame time of 16.67 ms, i.e. about 60 Hz or 60 frames/s.

The residual DC (RDC) property for photo-aligned LCD cells can be also measured. The RDC values can be measured using two methods: flicker-minimizing method and dc voltage method [64]. In the flicker-minimizing method, the test cell is driven by a square-wave 30Hz voltage with a varied DC offset, and the output signal with flicker could be observed [63]. The RDC is defined by the offset value, where minimum flicker is achieved.

DC voltage method can be also applied for the measurement. A voltage of 5V is applied to a cell for 10 minutes, and then the cell is shorted for 1 second. The RDC value is defined as the voltage measured after 10 minutes. The latter method is simpler but has less accuracy than the flicker-minimizing method. The acceptable value of RDC for TFT-LCD applications is less than 50mV [63].

Proper values of VHR and RDC requires a high purity of LC mixtures for TFT-LCD operations. The holding ratio must be high (>99%) and temperature-stable, which depends on the LC compounds used in AM-LCDs. This is possible only for well-purified LC mixtures containing fluorinated compounds (Figure 3.82) [1-3]. LCs with a large dielectric anisotropy are also difficult to apply, as the LC capacitance significantly decreases during the reorientation of the LC director in electrooptical effect, thus deteriorating the holding ratio.

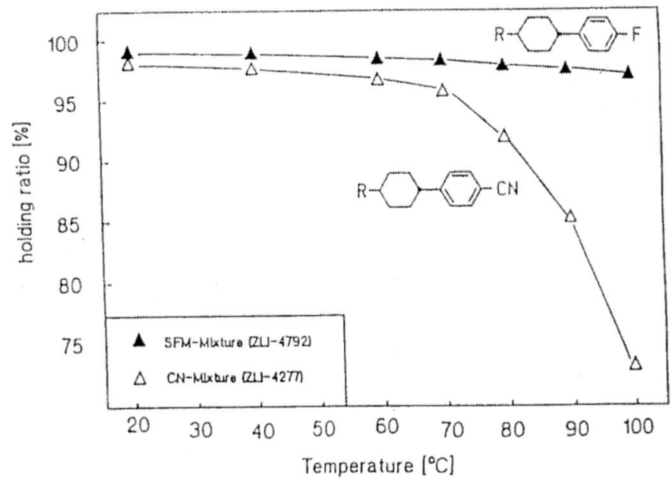

Figure 3.82. Holding ratio of TFT AM-LCD and its correlation with the LC molecular structure [1-3]. Temperature dependence of VHR for two different LC structures.

By all means the number of ions both in the bulk (evaluated by VHR) and at the surface of LC cell (evaluated by RDC) should ne minimized to provide an effective AM-LCD switching without flicker effect. The LC for AM-LCD should be of very high resistivity in order to hold the charges between frames. Actually LC switching off time:

$$\tau_{LC,\,off} \sim \varepsilon\rho, \tag{3.82}$$

where ε- is LC average dielectric constant (see 2.3.1) and ρ is a specific resistivity, that should be at least 10^{11} Ω·cm. For AM-LCD, LC with ρ of 10^{12} Ω·cm is not uncommon.

VHR and RDC values also correlate with the so-called “image-sticking “ effect, which takes place when LCD displayed the same image for sometime after the image was switched off [64]. The investigations in this field have shown that high VHR and low RDC prevents the image sticking effect in LCD [65]. The other factor, that correlates with an image sticking effect is LC anchoring energy, which must be sufficiently high to avoid the sticking phenomena [65].

AM-LCD are mostly based on the two electrooptical modes in LC: VAN mode, i.e. ECB effect in homeotropic LC cells (see also Figures 3.3, 3.5) and IPS (in-plane switching mode, Figure 3.83). The advantage of VAN mode is a very high contrast ratio, which is mostly defined by the quality of the polarizers and pretilt angle θ_0 (1-3^0 from the initial homeotropic alignment). IPS mode is characterized by a special geometry of electric field application parallel to the cell substrates (Figure 3.83). In this case the variation of the LC director polar angle θ and consequently the change of the LC cell phase retardation ΔΦ (3.15) is minimal and all the intensity changes are due to the LC twist angle φ_0 variations (3.16). As follows e.g. from (3.16) the intensity variations with the angle of light incidence are small in IPS mode, so IPS TFT-LCD viewing angles are very broad (Figure 3.83) [1,2].

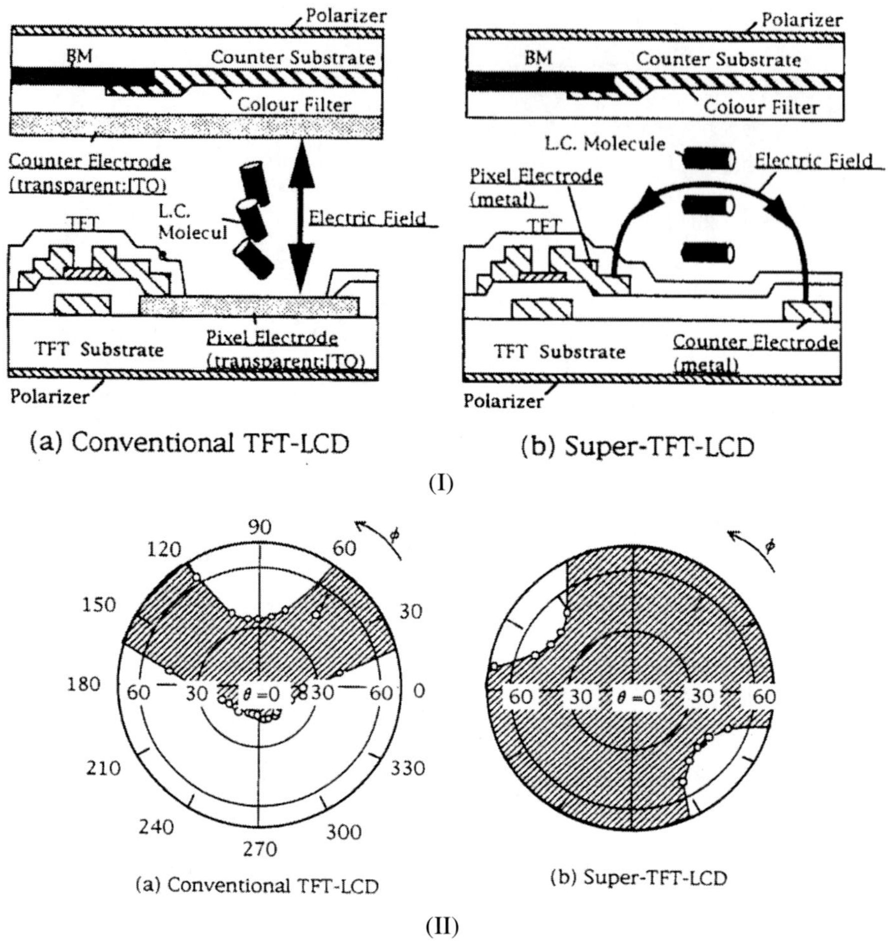

Figure 3.83. In-Plane-Switching Mode in Super TFT-LCD[1-3]. (I)- Geometry of realization of Super TFT-LCD; (II) –Comparison of viewing angle dependences with conventional TFT-LCD [1,3].

3.8.3. Addressing Schemes of Ferroelectric LC

The passive addressing of FLC displays is made similar to TN or STN –LCDs (Figure 3.75), i.e. select pulses address sequentially all the rows of the display matrix, while the data pulses address all the columns simultaneously in each select row time. As previously, the pixel voltage V_{ij} $=V_i$-V_j, i.e. the difference between a row V_i and column V_j voltage pulses
[1-3].

A basic difference between passive addressing of TN (STN) and FLC displays is the possibility to use an optical memory in the latter case. This means, that the sharp transmission –voltage curve (TVC) is no longer needed, as we do not have to refresh the voltage on the pixels. Ones switched the pixel of FLC screen will remain in this state. The problem is to make gray scale, because usually only two transmission levels are memorized. The gray levels can be achieved by rather artificial means either by time-control technique, i.e. making several sequential frame pictures for one image or space-control, i.e. dividing one pixel of the image to several sub-pixels, so that the transmission of the pixel can vary due to the number of switched sub-pixels [26]. Let us consider for simplicity FLC passive addressing scheme without gray scale. One of the schemes, called Seiko scheme [1-3] is shown above in Figure 3.74. Four voltage signals, so called *driving waveforms* are needed to address FLC matrix: two row signals, selecting V_s or non-selecting V_{ns} and two column signals controlled by the picture data (column white V_{on} and column black V_{off}). On the selected row the effective pixel voltages V_s-V_{on} and $V_s - V_{off}$ must produce switching of the pixel to one or another state within the selection time, which is called the line addressing time. At the same time the voltages $V_{ns} - V_{on}$ and $V_{ns} - V_{off}$ must not effect the state of the pixel. In Seiko scheme the requirements of zero average voltage over pixel is taken into account to avoid FLC degradation and the height of the writing pulses $U_R + U_D$ is three times as much as the height of AC stabilization pulses U_D. Here two scans are used to write one picture with a small interrupting time between them- one subframe to write white pixels and another to write black ones. Total time to write one row in Seiko scheme is $N\tau$, where τ is the time required to achieve one or another bistable FLC state using the pulses with the amplitude of $U_R + U_D$ and N=2+2=4 –a number of intervals. The scheme is not optimized as the value of N is too large. Minimization of the value of N is the purpose of the newly developed FLC addressing schemes [1-3].

The multiplexing ratio M of passively addressed FLC display is defined by the frame time τ_{fr}, and the row writing time (or line address time) $N\tau$ according to the formula:

$$M=\tau_{fr}/N\tau \tag{3.83}$$

The addressing of FLC displays is the example of addressing LCD with memory effect. The most important parameter, which define the resolution (number of lines N) in this case is the addressing time τ, which is required for the fixation of the memory effect.

REFERENCES

[1] Chigrinov, VG. Liquid Crystal Devices: Physics and Applications, (Artech-House, 1999).
[2] Blinov, LM; Chigrinov, VG. Electrooptical effects in liquid crystalline materials, (Springer-Verlag, 1994).
[3] Pasechnik, SV; Chigrinov, VG; Shmeliova, DV. Liquid Crystals, Viscous and Elastic Properties, (Wiley-VCH, Berlin, 2009).
[4] Vladimir Chigrinov, Dmitry Yakovlev, and Hoi Sing Kwok, Optimization and Modeling of Liquid Crystal Displays, Information Display, 20, 26 (2004).
[5] Yeh, P; Gu, C; *Optics of Liquid Crystal Displays*, (Wiley, 1999).
[6] H. Wohler, and M.E. Becker, *Opto-Electronics review*, 10, 23 (2002).
[7] http://www.lc-dimos.com/
[8] http://www.shintech.jp/
[9] http://www.sanayisystem.com

[10] Kohki Takatoh, Mitsuhiro Akimoto, Hiroomi Kaneko, Koji Kawashima, Shunsuke Kobayashi, The Low Driving Voltage, the Molecular Arrangement and the Stability of Reverse TN mode, *IDW'08 Digest*, p. 1567 (2008).

[11] Li, Y.W; Tan, L; Kwok, HS. Field-Sequential-Color LCDs Based on Transient Modes, *SID'08 Digest*, p. 32 (2008).

[12] Scheffer, TJ; Nehring, J. A new, highly multiplexable liquid crystal display, *Appl. Phys. Lett.*, 45, 1021 (1984).

[13] Schadt, M; Leenhouts, F. Electro-optical performance of a new, black-white and highly multiplexable liquid crystal display, *Appl. Phys. Lett.*, 50, 236 (1987).

[14] Scheffer, TJ; Nehring, J. Twisted Nematic and Supertwisted Nematic Mode LCDs, In Liquid Crystals: Applications and Uses, (edited by B.Bahadur), World Scientific, 1991.

[15] Li, J; Kelly, J; Hoke, Ch. Bos, PJ. Optimization of luminance for optical-mode-interference (OMI) displays, *Journal of the SID*, 4/2, 95 (1996).

[16] Tang, ST; Kwok, HS. New method of designing LCD with optimal optical properties, SID'99 Digest (1999).

[17] http://www.ece.ust.hk/~eekwok/publications/1999/ASID5-1999-165.pdf

[18] Mori, H; Nagai, M; Nakayama, H; Itoh, Y; Kamada, K; Arakawa K; Kawata, K. Novel Optical Compensation Method Based upon a Discotic Optical Compensation Film for Wide-Viewing-Angle LCDs, *SID'2003 Digest*, p. 1058 (2003).

[19] Yang, DK; Wu, ST. Fundamentals of Liquid Crystal Devices, Wiley, 2006.

[20] Nithya Venkataraman, Greg Magyar, Erica Montbach, Asad Khan, Tod Schneider, J. William Doane, Lisa Green, and Quan Li, Thin Flexible Photosensitive Cholesteric Displays, *IDRC'08 Digest*, p. 101 (2008).

[21] Garoff, S; Meyer, B. Electroclinic effect at the A-C phase change in a chiral smectic liquid crystal, *Phys. Rev. A.*, 19, 338 (1979).

[22] Pozhidaev, EP; Osipov, MA; Chigrinov, VG; Baikalov, VA; Blinov, LM; Beresnev, LA. Rotational viscosity of smectic C* phase of ferroelectric liquid crystals, *Soviet Physics - JETP*, 67, 283 (1988).

[23] Chilaya, GS; Chigrinov, VG. Optics and electrooptics of chiral smectic C liquid crystals, *Physics-Uspekhi*, 36, 909 (1993).

[24] Chigrinov, V; Panarin, Yu. Vorflusev, V; Pozhidaev, E. Aligning properties and anchoring strength of ferroelectric liquid crystals, *Ferroelectrics*, 178, 145 (1996).

[25] Clark, NA; Lagerwall, ST. Submicrosecond bistable electrooptic switching in liquid crystals, *Appl. Phys. Lett.*, 36, 899 (1980).

[26] Lagerwall, ST. Ferroelectric and antiferroelectric liquid crystals, *Ferroelectrics*, 301, 15, (2004).

[27] Eugene Pozhidaev, Vladimir Chigrinov, Gurumurthy Hegde, Peizhi Xu, Multistable electro-optical modes in ferroelectric liquid crystals, *J. SID*, 17/1, 53 (2009).

[28] Pozhidaev, E.P; Chigrinov, V.G. Bistable and Multistable States in Ferroelectric Liquid Crystals, *Cryst. Reps.*, 51, 1030 (2006).

[29] Beresnev, L.A; Chigrinov, V.G.; Dergachev, D.I.; Pozhidaev, E.P.; Funfschilling, J; Schadt, M. Deformed helix ferroelectric liquid crystal display: a new electrooptic mode in ferroelectric chiral smectic C liquid crystal, *Liq. Cryst.*, 5, 1171 (1989).

[30] Panarin, Yu; Pozhidaev, E; Chigrinov, V. Dynamics of controlled birefringence in an electric fielf deformed helical structure of a ferroelectric liquid crystal, *Ferroelectrics*, 114, 181 (1991).

[31] Pozhidaev, EP. *Electrooptical properties of deformed-helix ferroelectric liquid crystal display cells*, SPIE Digest, *4511*, 92 (2001).
[32] Vladimir Presnyakov, Zhijian Liu, Vladimir G. Chigrinov, Fast optical retarder using deformed-helical ferroelectric liquid crystals, *Proc. SPIE*, 5970, 426 (2005).
[33] Evgeny Pozhidaev, Vladimir Chigrinov, Fast Photo-Aligned V-shape Ferroelectric LCD Based on DHF Mode, SID'10 Digest, p. 387, (2010).
[34] Gurumurthy Hegde, Peizhi Xu, Eugene Pozhidaev, Vladimir Chigrinov, Hoi Sing Kwok, Electrically controlled birefringence colours in deformed helix ferroelectric liquid crystals, *Liq. Cryst*, 35, 1137 (2008).
[35] Vladimir G. Chigrinov and Hoi-Sing Kwok, Photoalignment of liquid crystals: applications to fast response ferroelectric liquid crystals and rewritable photonic devices, In a book: "Progress in Liquid Crystal Science and Technology: in Honor of Shunsuke Kobayashi's 80th Birthday", World Scientific, February 2013.
[36] Dmitry A. Yakovlev, Vladimir G. Chigrinov, Hoi-Sing Kwok, Modeling and Optimization of LCD Optical Performance (Wiley Series in Display Technology), 576 pp, Wiley-Blackwell; 1 edition (3 Oct 2014).
[37] Alexei D. Kiselev, Eugene P. Pozhidaev, Vladimir G. Chigrinov, Hoi-Sing Kwok, Polarization-gratings approach to deformed-helix ferroelectric liquid crystals with subwavelength pitch, *Phys. Rev. E*, 83, 031703 (1-11), 2011.
[38] Pozhidaev, E.P.; Minchenko, M; Molkin, V; Torgova, S; Srivastava, A.K.; Chigrinov, V; Kwok, H-S; Vashenko, V; Krivoshey, A. High Frequency Low Voltage Shock-Free Ferroelectric Liquid Crystal: A New Electro-Optical Mode with Electrically Suppressed Helix, *Oral* 1.3, Eurodisplay 2011, Arcachon, France, September, 2011.
[39] Pozhidaev, E.P.; Kiselev, AD; Chigrinov, V.G.; Srivastava, A.K.; Minchenko, M.V.; Kotova, S.P., Orientational Kerr-effect in Ferroelectric Liquid Crystals, *Abstracts of 24th Intern. Liq. Cryst.* Conf, Mainz, Germany, August 2012.
[40] Evgeny P. Pozhidaev, Alexei D. Kiselev, Abhishek Kumar Srivastava,Vladimir G. Chigrinov, Hoi-Sing Kwok, Maxim V. Minchenko, Orientational Kerr effect and phase modulation of light in deformed-helix ferroelectric liquid crystals with subwavelength pitch, *Phys. Rev. E*, 87, 052502 (1-8), 2013.
[41] Rao, L; Yan, J; Wu, ST; Yamamoto, S; Haseba, Y. Polymer-stabilized blue phase liquid crystal with a negative Kerr constant, *Appl. Phys. Lett.*, 98, 081109 (2010).
[42] Angelé, J; Elyaakoubi, M; Guignard, F; Jacquier, S; Joly, S; Martinot-Lagarde, P; Osterman, J; Laffitte, JD; Leblanc, F. An A4 BiNem Displays with 3.8 Mega Pixels, *IDW'07 Digest*, p. 265 (2007).
[43] Yuet-Wing Li, Chung Yung Lee Hoi Sing Kwok, Permanent Bistable Twisted Nematic Displays using Bi-directional Alignment Surface, *SID'08 Digest*, p. 1026 (2008).
[44] Cliff Jones, J. The Zenithal Bistable Device: From Concept to Consumer, *SID'07 Digest*, p. 1347 (2007).
[45] Alexander Muravsky, Anatoli Murauski, Vladimir Chigrinov, Hoi-Sing Kwok, Optical Rewritable Electronic Paper, IEICE TRANS. *ELECTRON.*, E91–C, 1576 (2008).
[46] Chigrinov, VG; Kozenkov, VM; Kwok, HS. Photoalignment of Liquid Crystalline Materials: Physics and Applications, Wiley, (2008).
[47] Ko-Ting Cheng, Cheng-Kai Liu, Chi-Lun Ting, Fuh AY-G, Optical addressing in dye-doped cholesteric liquid crystals, *Opt. Communs.*, 281, 5133 (2008).

[48] Masahiro Sato, Tsutomu Ishii, Naoki Hiji,Kyotaro Tomoda, Shigeru Yamamoto, and Kazuo Baba, High Resolution Electronic Paper Based on LED Print Head Scanning Exposure, *SID'08 Digest*, p.923 (2008).

[49] Wei Hu, Abhishek Srivastava, Fei Xu, Jia-Tong Sun, Xiao-Wen Lin, Hong-Qing Cui, Vladimir Chigrinov and Yan-Qing Lu1, Liquid crystal gratings based on alternate TN and PA photoalignment, *Optics Express*, Vol. 20, No. 5, 5384-5391 (2012).

[50] Wei Hu, Abhishek Kumar Srivastava, Xiao-Wen Lin, Xiao Liang, Zi-Jian Wu, Jia-Tong Sun, Ge Zhu, Vladimir Chigrinov, and Yan-Qing Lu, Polarization independent liquid crystal gratings based on orthogonal photoalignments, *Appl. Phys. Lett.*, 100, 111116-1-3 (2012).

[51] Yan-qing Lu, Wei Hu, Xiao-wen Lin, Abhishek Srivastava, Vladimir G. Chigrinov, Liquid crystal gratings, from nematic to blue phase, SPIE Optics+ Photonics, San Diego, August, 2012.

[52] Xiao-Wen Lin, Wei Hu, Xi-Kui Hu, Xiao Liang, Ye Chen, Hong-Qing Cui, Ge Zhu, Jia-Nan Li, Vladimir Chigrinov, and Yan-Qing Lu, Fast response dual-frequency liquid crystal switch with photo-patterned alignments, *Opt. Letts*, 37, No. 18, 1-3 (2012).

[53] Stephen M. Morris, Damian J. Gardiner, Flynn Castles, Philip J. W. Hands, Timothy D. Wilkinson, and Harry J. Coles, Fast-switching phase gratings using in-plane addressed short-pitch polymer stabilized chiral nematic liquid crystals, *Appl. Phys. Lett.*, 99, 253502-1-3 (2011).

[54] Carbone, G; Salter, P; Elston, SJ; Raynes, P; De Sio, L; Ferjani, S; Strangi, G; Umeton, C; Bartolino, R. ,Short pitch cholesteric electro-optical device based on periodic polymer structures, *Appl. Phys. Lett.*, 95, 011102-1-3 (2009).

[55] Fan, F; Srivastava, AK; Chigrinov, VG; Kwok, HS., Switchable liquid crystal grating with sub millisecond response, *Appl. Phys. Lett.*, 100, 111105-1-3 (2012).

[56] Srivastava, AK; Wei Hu, Chigrinov, VG; Kiselev, AD; Yan-Qing Lu, Fast switchable grating based on orthogonal photo alignments of ferroelectric liquid crystals, *Appl. Phys. Lett.*, 101, 031112-1-4 (2012).

[57] Date, M ; Naito, N; Tanaka, K; Kato, K; Sakai, S., Three-Primary-Color Holographic Polymer Dispersed Liquid Crystal (HPDLC) Devices for Reflective Displays, *Proceedings of Asia Display'95*, 1995, 803-806.

[58] Tanaka, K; Kato, K; Date, M; Sakai, S., Optimization of Holographic PDLC for Reflective Color Display Applications, *SID'95 Digest*, 1995, pp.267-270.

[59] Crawford, GP. Liquid Crystal PolymerDispersions for Reflective Flat-Panel Displays, *SID'96 Digest*, 1996, F-4/1-F/4-47.

[60] Zhigang Zheng, Jing Song, Yonggang Liu, Fuzhong Guo, Ji Ma, Li Xuan, Single-step exposure for two-dimensional electrically-tuneable diffraction grating based on polymer dispersed liquid crystal, *Liq. Cryst.*, 35, 489 (2008).

[61] Wu, ST; Fuh, AY. Two-Dimensional Diffraction Grating Based on Polymer-Dispersed Liquid Crystals, *Jpn. J. Appl. Phys.* 43, 7077–7082 (2004).

[62] Liu, YJ; Sun, XW; Shum, P; Yin, XJ. Tunable fly's-eye lens made of patterned polymer-dispersed liquid crystal, *Opt. Express*, 14, 5634–5640 (2006).

[63] Huang, HC; Cheng, PW; Kwok, HS. On the Minimization of Flicker and Image Retention in Silicon Light Valves, *SID'00 Digest*, p.248 (2000).

[64] Nakcho Choi, Hyunku Ahn and Sungtae Shin, Developing the new Evaluation Method of the Image Sticking, IDW'08 Digest, p. 43 (2008).

[65] Woo, JW; Shin, DC; Lim, EJ; Park, HJ; Shin, HH. UV photoalignment technology comparable to the rubbing alignment method in terms of the azimuthal anchoring energy, *IDW'08 Digest*, p. 37 (2008).

Chapter 4

LIQUID CRYSTAL PHOTONICS

4.1. LC OPTICAL ELEMENTS FOR FIBER OPTICAL COMMUNICATION SYSTEMS

Photonics LC applications include passive LC elements for fiber optical communication systems (DWDM components) based on liquid crystal cells, which can successfully compete with the other elements used for the purpose, such as micro electromechanical (MEM), thermo-optical, opto-mechanical or acousto-optical devices [1]. The known LC applications in fiber optics enable to produce switches, filters, attenuators, equalizers, polarization controllers, phase emulators and other fiber optical components. Application of nematic and ferroelectric LC for high speed communication systems, producing elements that are extremely fast, stable, durable, of low loss, operable over a wide temperature range, and that require low operating voltages and extremely low power consumption. Good robustness due to the absence of moving parts and compatibility with VLSI technology, excellent parameters in a large photonic wavelength range, whereas the complexity of the design and the cost of the device are equivalent to regular passive matrix LC displays makes LC fiber optical devices very attractive for mass production. The quality criteria for photonics fiber optical LC applications may be also very different for various applications. For instance the LC Variable optical attenuators (VOA) are characterized by: (i) crosstalk attenuation; (ii) insertion loss; (iii) operation wavelength; (iv) working temperature range; (v) switching power; (vi) response time. The photonics LC devices should be standardized for the same criteria as usual fiber optical components for photonics applications.

FTTH (fiber- to-the-home) is a form of fiber optic communication delivery in which the fiber extends from the central office to the subscriber's living or working space [2]. Fiber configurations that bring fiber right into the building can offer the highest speeds delivering voice, video and data since the Gigabit and Terabit Ethernet can be efficiently used. FTTH will require a number of passive optical elements, such as switches, filters, attenuators, polarization rotators and controllers, which can be efficiently implemented using novel LC technology. Fast switching LC cells with microsecond and sub-microsecond switching time should replace the currently used micro-electro-mechanical (MEM) switching devices with millisecond switching time presently used in FTTH systems.

Despite a certain drop in the fiber optical component market within the last three years, the need for reliable passive optical components is still very strong and will continue to grow in the long run. Certain LC components, such as LC based coaxial variable optical attenuators, polarization controllers and phase retarders have already appeared in the market [1-4]. Recently, Vescent Photonics (USA) announced the LC waveguides as a new electro-optic technology platform for a variety of applications, such as interferometers, beam steerers, tunable filters and lasers etc. [4]. We will consider various photonics applications of LC devices.

4.1.1. Switches

Switches for optical fiber optical networks are increasingly important. LC switches show certain advantages in comparison with MEM switches, commonly used for the same purpose, such as (i) fast switching time; (ii) low controlling voltages and power consumption; (iii) higher reliability and working time [1]. However, wavelength dependence of the response times and thermal drift of the characteristics of LC switches should be avoided [1]. There are two main techniques, which can be used in LC optical switches, working in non-polarized light. In the first of them the non-polarized light from the input fiber is decomposed by polarizing beam splitters (PBS) on the two orthogonal polarizations, which are then independently rotated by the SLM and collected on the output fiber, thus transferring the input non-polarized light to the proper output channel. The fast switching of optical channels in nonpolarized light was proposed in [5] (Figure 4.1).

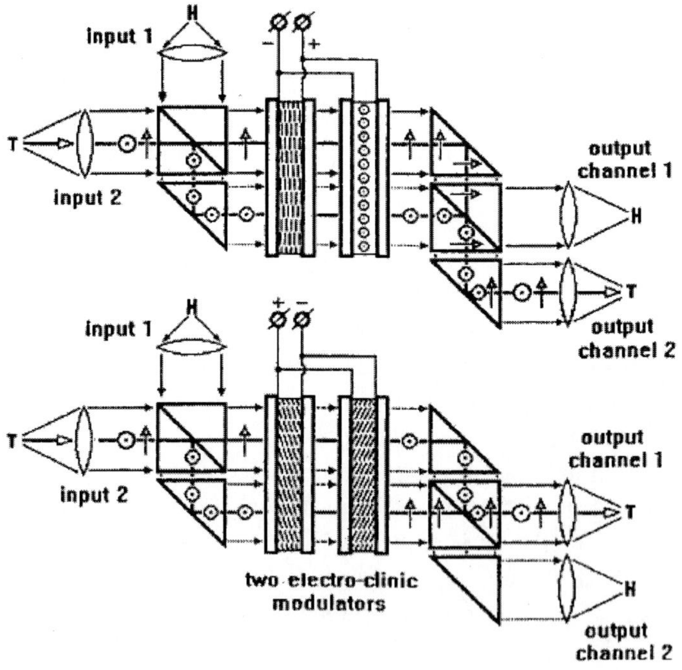

Figure 4.1. FLC switch of nonpolarized light images between two input and two output channels based on two polarization beam splitters and two FLC electroclinic modulators, mounted in series. The transfer of the images (letters 'H' and 'T') is shown in two different cases [5].

The switching device includes ferroelectric LC modulators and polarization beam splitters, which may control the polarization of four polarized input beams, transferring the input images (letters 'H' and 'T') from the input 1 or 2 to the corresponding outputs (channels 1 and 2). The transfer of the images 1 => 1 and 2 => 2 or 1 => 2 and 2 => 1 depends on the amplitude and sign of the voltage applied to the modulators (Figure 4.1). The advantage of the switch is very fast response time (5 μsec at the voltage ±30V). The device has an intrinsic grayscale and very small wavelength dispersion, making it possible to redistribute the input light intensities between the outputs within the whole visible range. An integrated LC switch has been also proposed and designed that works independently of the polarization state of the input light [6]. The proposed structure consisted of a polarization splitter, two polarization converters, and a polarization combiner. The light propagation patterns within the whole device in the OFF and ON states for both TE and TM modes have been demonstrated.

In the second the voltage controllable diffraction can be used for the optical switch [7, 8]. The LC optical switching can be also arranged by changing the boundary refractive index for the fiber using LC cell [9,10]. Another LC optical switches can be used utilizing the principle of Mach Zehnder interferometer with a switchable ferroelectric LC cladding [11] or Fabry-Perot LC infiltration cell integrated in high index contrast silicon-on-insulator waveguide [12].

The following electrooptical effects in liquid crystals can be used in optical switching.

1. Electroclinic effect in smectic LCs [5] (Figure 4.1). Electroclinic effect is a phase transition between Smectic A (Figure 2.4, b) and Smectic C phase (Figure 2.4, c) arranged by electric field [13,14]. The switching time of 5 microseconds was reported for the controlling voltages of ±30V (Figure 4.1) [5]. The effect has a weak dependence on the light wavelength. However, the switching must be provided in the double cell configuration (one cell does not allow ninety degree rotation of the light polarization).
2. Deformed Helix Ferrroelectric (DHF) Effect in Ferroelectric LCs (FLCs) (Figure 3.40). The switching time, less than 10 μs at the controlling voltage of ±20V can be provided, which is temperature independent over the broad temperature range [15]. The DHF LC cells are the basis of fast responding modulators and optical filters, which are applicable in fiber optical communication systems. The fast ferroelectric LC shutters (deformed helix ferroelectric-DHF effect) with the response time less than one microsecond in a broad temperature range from 20 to 80 degrees were also developed [16] (Figures 3.41, 3.42). We believe, that DHF FLC is the fastest electrooptical mode in LC cells for Photonics applications.
3. Nematic LC cells with 270 degrees twist can provide the switching times less than 5 ms [13,14]. The prototypes of LC modulators are already known. The modulators can be used as LC fiber optical switch. The LC switches can use the effect of total internal reflection in nematic LC [17,18] or selective reflection effect in cholesteric LC [19]. The total internal reflection switch operates only for one light polarization (TE mode) and the most promising is VAN configuration (Figure 3.1,b) [18]. The switching time of 1 ms can be easily obtained in this case for the switching pulse amplitude of 5V (Figure 4.2) [18].

4. Bistable nematic switches are also possible, using bistable nematic twisted (BTN) cells [20]. In the latter case, the switching power is reduced, as the LC cell keeps the switched state without voltage.
5. The bypass optical switch based on two nematic liquid crystal cells with a switching time less than 200μs was demonstrated using two temperature-stabilized nematic LC birefringent cells [21] (Figure 4.3). Two subsequent NLC cells with crossed optical axes compensate the relaxation of NLC birefringence if turned off simultaneously. Thus the switching speed of two-cell switch can be as fast as NLC cell turn-on times. The cells can be specified to a certain fiber wavelength by adjusting the cell gap thickness.
6. LC switch can be made to control light beams in a plane of LC layers (Figure 4.4) [22]. It was shown experimentally that the direction of the light beam can be considerably changed by refraction and reflection of light at the sharp boundaries between the regions with different orientations [22]. LC switching can be controlled by electric field. Certain ways were proposed for optimization insertion loss and crosstalk of the 1x2 switch for real photonic applications (Figure 4.4).

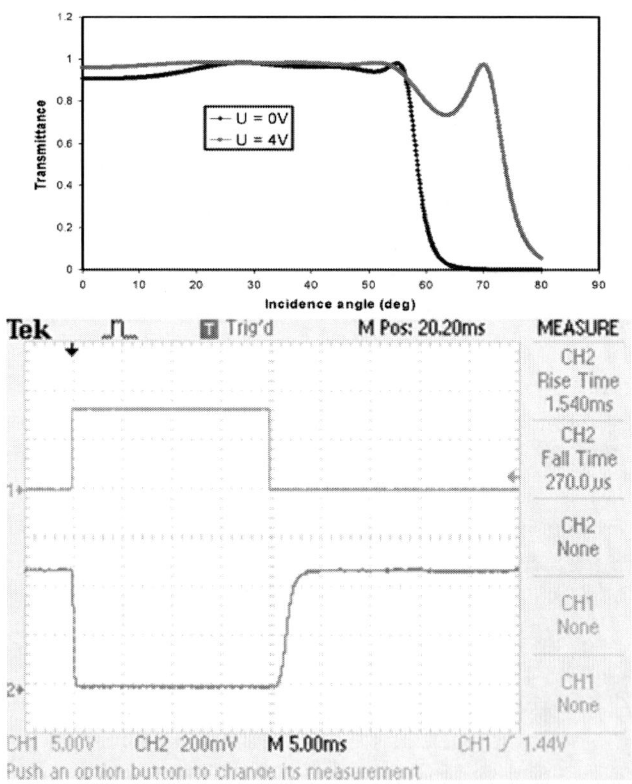

Figure 4.2. Total internal reflection (TIR) in nematic LC [18]. Left: calculations of TIR transmission versus light incidence angle for two voltages U =0V and U=4V. Right: experimental data of switching dynamics of TE mode in VAN configuration (Figure 3.1, b).

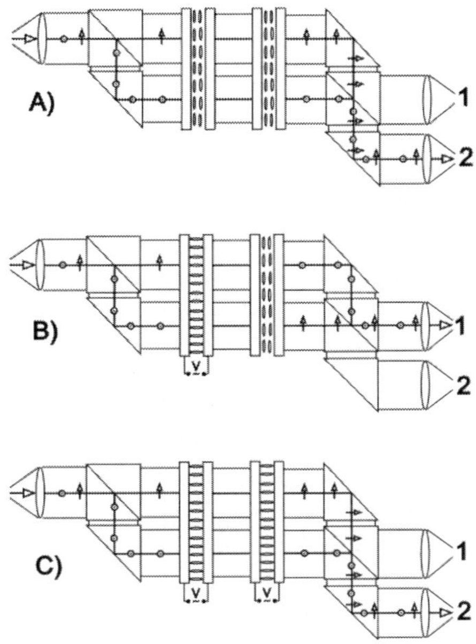

Figure 4.3. A bypass two-cell optical switch based on nematic liquid crystal [21].

Using different ITO templates, it is possible to create NxM switch and other different optical processing data elements, e.g. attenuators. There are a number of ways to optimize such types of LC devices including application of fast operating ferroelectric liquid crystal layers, which can provide the operation times in a microsecond range.

Polarization controllers are the elements that can transfer any input state of polarization (SOP) to the desired one, thus controlling the unpredictable polarization change or drift, which comes from the polarization-dependent components of the fiber optical system. These elements can be made on the basis of the three subsequently placed LC cells, which exhibit the effect of electrically controlled birefringence (ECB) with a homogeneous initial orientation [23]. Typical switching times are dependent on LC material and of the order of 10ms for the wavelength λ=1.3μ [23] (Figure 4.5).

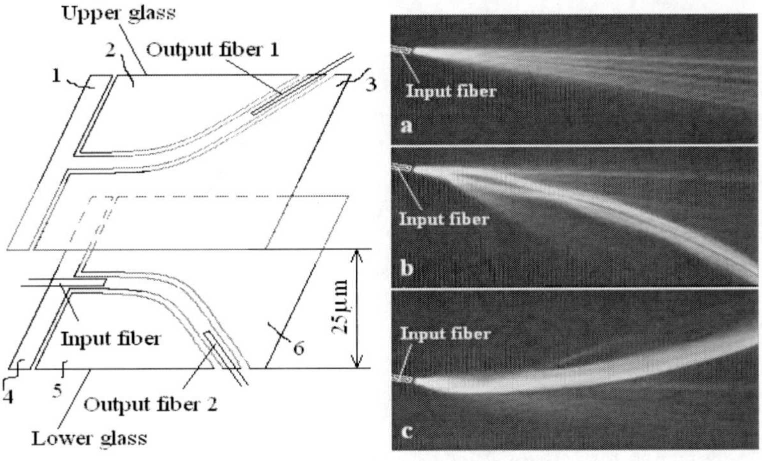

Figure 4.4. A 1x2 optical switch based on waveguide effect in photoaligned LC layer [22].

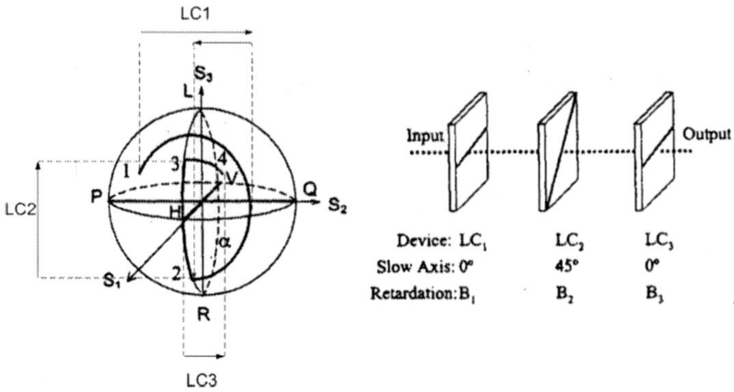

Figure 4.5. A polarization controller constructed with a cascade of three LC cells [23]. This configuration allows a fast arbitrary-to-arbitrary transformation of state of polarization (SOP) on the Poincare Sphere (left). The diagram on the left shows a transformation of SOP at point 1 to a final SOP at point 4 following two intermediate transformation at point 2 and point 3 [23].

4.1.2. Polarization Rotators

Polarization rotator is an LC element, that can rotate the linear polarized light in the output of the LC system to any desirable angle still keeping the linear polarization at the output [24, 25]. A novel configuration for a TN-LCD between two quarter-wave plates can operate as a controllable polarization rotator [24]. The plane of polarization of the transmitted light can be rotated as a function of the phase introduced by TN-LCD for different voltages [24]. The other LC configuration consists of the polarizer and two LC homogeneous cells placed at 45^0 with respect to each other (one with the a voltage controllable phase change and the other fixed as a quarter wave plate, Figure 4.6) [25]. The configuration can rotate the light polarization state at any angle between 0^0 and 90^0 dependent on voltage.

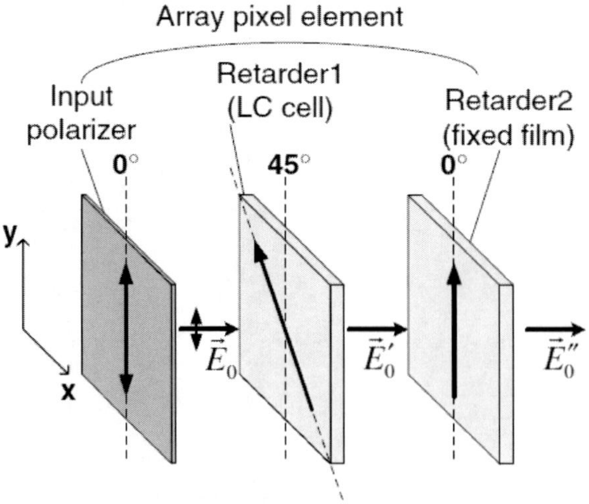

Figure 4.6. LC polarization rotator, consisted of a polarizer and two phase retarders, one of which is controlled by a voltage. The rotation of the polarization plane is between 0^0 and 90^0 [25].

4.1.3. Variable Optical Attenuators

Variable optical LC attenuators (VOA) have the typical attenuation range of 30 dB for the driving voltage of 12V at 1525-1575 wavelength range [26-29] with the response time of about 10-30 ms. Some of them are based on the light scattering of LC cell filled with a polymer network (PN-LCD) [13] (Figure 4.7). The light from the input fiber was scattered as a result of refractive index mismatching of the LC and the polymer when no voltage was applied, and it was passed through the PN-LC layer as a result of the refractive index matching of the LC and the polymer, when a voltage was applied (Figure 4.7).

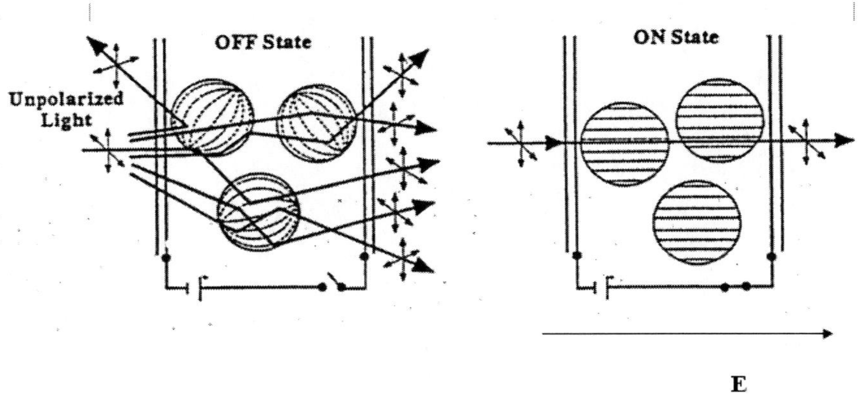

Figure 4.7. The structure and electrooptic properties of PN-LC films, attenuated non-polarized light. Left: field off state, scattering unpolarized light; right-transparent for the normally incident light field on state [13].

4.1.4. Photonic Crystal/ Liquid Crystal Structures

Another promising technique for making passive fiber optics components is the use of photonic crystal (PC) materials [1]. Photonic crystals are artificial media with 1D, 2D or 3D periodical structures that exhibit both unidirectional and omnidirectional photonic band gaps. The band gap defines a range of frequencies at which electromagnetic radiation striking the crystal is reflected by the crystal rather than being permitted to propagate through it. In these crystals, the dimensions of the lattice structures and the dielectric elements are selected to produce band gaps having desired center frequencies and bandwidths. Electromagnetic radiation at a frequency within the band gap is reflected from the structure via the well-known Bragg reflection phenomenon. Electrically tunable photonic band gap (PBG) structures hold the potential to become a versatile and compact backbone for optical signal processing. Electrical tuning of two-dimensional PBG structures infiltrated with liquid crystals was reported [30-34]. The structures can be operated with any electro-optic materials and lead to fast and efficient modulators, routers, and tunable filters [30]. The effect of LC alignment is very important, in particular, a high quality LC alignment parallel to the photonic hole surface is highly desirable [1]. The example of voltage controllable photonic crystal/ liquid crystal (PC/LC) structure is given in Figure 4.8 [33]. Liquid crystals are materials that present a variation in their effective refractive index if an electric field is applied (3.14). In addition to this LC are cost effective and easy to manipulate [13]; that is the reason for selecting this material instead of others dependent on the temperature or a mechanical force for instance. By changing the voltage applied to the structure, the defect state that constitutes the filter will sweep a range of wavelengths [33]. Other important applications of liquid crystals in the middle of one-dimensional PC/LC structures are electro-optical switches and modulators [34].

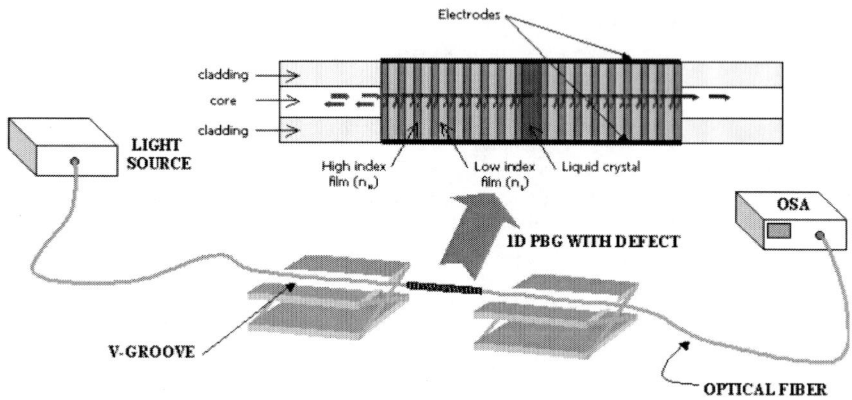

Figure 4.8. Structure of the 1D PBG wavelength filter consisting of 2 Bragg mirrors of 30 layers and a cavity (defect) [33].

Controllable photonic crystal / liquid crystal (PC/LC) devices became a “hot” topic of research. Photonic crystals are known to provide wavelength-dependent filters, beam splitters and mirror components. As photonic crystals are on the order of several wavelengths, in each of two major dimensions, and are made with microprocessing techniques, optical processing systems employing photonic crystals can be very small, thereby permitting extremely high bit density and high rates of data processing. Filling the interstices of the photonic crystal with the properly aligned LC material, and subjecting the LC to a varying electric field can produce a tunable photonic crystal element. We will consider below the photoaligning materials to align LC mixtures in small cavities, such as the holes and tubes of photonic crystals, having size of 1 μm and less and obtained excellent LC orientation inside the tubes by photoalignment. New PC/LC passive elements of fiber optical systems include optical switches, fiber optical isolators, optical filters, wavelength selective mirrors for fiber lasers, polarization controllers, attenuators etc., [1].

4.2. Photoalignment Technology for LC Photonics Devices

Photo-alignment possesses obvious advantages in comparison with the usually “rubbing” treatment of the substrates of liquid crystal display (LCD) cells. Possible benefits for using this technique in Photonics LC devices include (see also Fig. 2.18) [35]:

(i) new advanced applications of LC in fiber communications, optical data processing, holography and other fields, where the traditional rubbing LC alignment is not possible due to the sophisticated geometry of LC cell and/or high spatial resolution of the processing system;

(ii) ability for efficient LC alignment on curved and flexible substrates;

(iii) manufacturing of new optical elements for LC technology, such as patterned polarizers and phase retarders, tunable optical filters, variable optical attenuators etc.

The photoaligning materials can be based on photo-polymerized and cross-linked dye photosensitive layers, that enable [35] (i) high order parameter more than 0.8; (ii) excellent alignment quality of nematic and ferroelectric LC materials in various modes; (iii) temperature and UV stability due to the polymerization and cross-linking effect in dye layers; (iv) perfect adhesion and anchoring energy comparable with rubbed polyimide (PI) layers; (vi) excellent sensitivity with a minimum exposure energy; (vii) ability to align LC materials in curved surfaces and photonic holes.

Photonic crystal fiber is a glass or polymer fiber with an array of microscopic air holes running along the length of the fiber. The waveguide properties of such a fiber can be controlled by introducing an additional material into the air holes [30-32]. LC is suitable for that purpose because its refractive index can be easily tuned by electric field or temperature. The technique of photo-configurable alignment of LC in glass micro-tubes and in photonic crystal fiber was developed (Figure 4.9) [36,37]. Good homogeneous alignment was detected with polarized microscopy and FTIR spectroscopy methods. The presented technique of alignment is based on proper developed photoaligning azo-dye material and is promising as a non-contact method for LC orientation in complex photonic crystal structures [36]. Figure 4.9 presents the glass tube of inner diameter 4 μm treated with the photoaligning sulfonic azo-layer (SD1) [35] and filled with uniform nematic LC orientation without point defects or linear disclinations. The order parameter S of LC has been obtained from FTIR spectroscopy data and has demonstrated good alignment quality (S=0.63 [36]). Presented technique can be used as non-contact method of LC alignment in complex photonic crystal structure.

An effective method of photoaligning of LC molecules in silica micro-capillaries with the use of the SD-1 azo dyes has been developed [37]. The presented method has a long list of advantages in the context of applications for photonic liquid crystal fibers: no need for direct access for an aligned surface, ability to obtain different alignment with a single material, very good and stable orientation, and finally, the possibility to obtain patterned alignments within a single micro-capillary. The simplest and fastest method is filling without any pre-treatment of capillaries; however, the quality of such samples is sufficient only for some basic and fast experimental investigations (i.e. where long-term stability and low scattering losses are not crucial [37]). The photoalignment technique is most sophisticated, but it is the only technique that allows to obtain different alignment in a photonic LC fibers with a single technological process, and moreover some alignments are possible to obtain only with this method (i.e. transverse, transverse tilted and patterned alignments). The application of the presented technique for creating well aligned LC in fiber can lead to a variety of interesting new phenomena. It can be expected that fibers with a transverse molecular orientation would be characterized by relatively high birefringence, which can be tuned with an external electric field and the effect of tuning would be strongly dependent on the direction of an electric field with respect to the photo-induced birefringence axis. The efficiency of tuning would be higher if a transverse-tilted orientation were generated, especially if a dual frequency LC were used. Periodically patterned alignment can be also very interesting in the context of LC fibers, since it can be used for creating a new type of long-period fibre gratings [37]. Moreover, it can be also expected that other interesting alignment textures can be obtained if a more sophisticated irradiation scheme is used (Figure 4.10) [37].

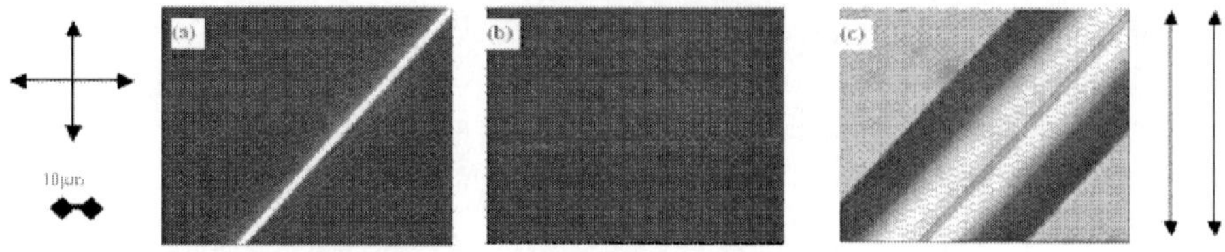

Figure 4.9. Photo-alignment in micro-tube [36], crossed polarizers: (a) angle between polarizer and tubes axis is 45 degrees; (b) angle between polarizer and tubes axis is 0 degrees; parallel polarizers: (c) -angle between polarizer and tubes axis is 45 degrees [36,37].

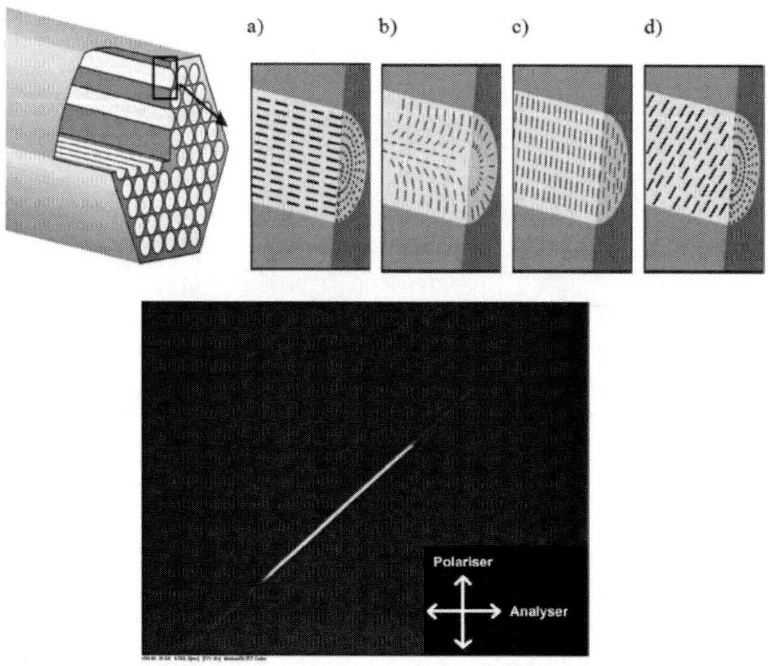

Figure 4.10. Photo-alignment in photonic LC fiber [37]. Above: types of LC alignment; Below: Single-frame except from video recording of micro-capillaries filled with LC, in which two stable alignments (planar and tilted at 45°) have been obtained simultaneously.

The uniform azo-dye photoalignment on the profiled 3D surface (substrate with bulk relief) was also demonstrated [38]. The three step exposure process for uniform surface alignment was developed (Figure 4.11). Patterned exposure of azo-materials is a very useful procedure for LC photonic devices as a high quality LC alignment on profiled surfaces can be obtained.

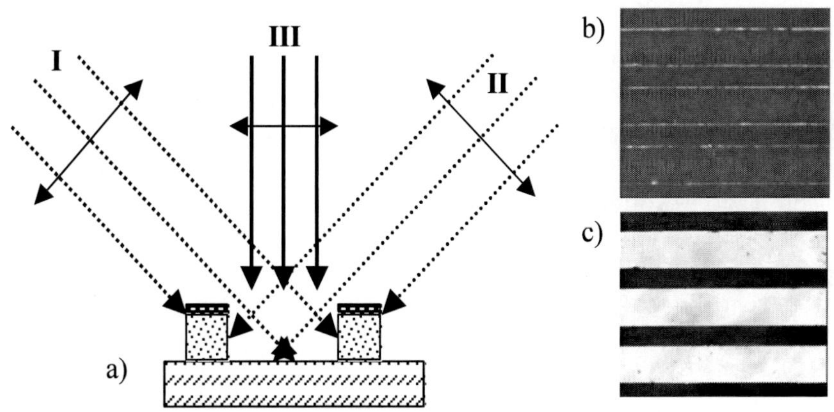

Figure 4.11. Optical scheme of three step 3D-substrate illumination (a); Profiled substrate LC cell between crossed (b) and parallel (c) polarizers [38].

LC alignment on submicrometer-sized rib waveguides on silicon chips was studied [1, 35]. The experiments using nematic liquid crystal cladding on silicon waveguides and microrings coated with photoalignment layer, and covered with a vertically aligned polyimide rubbed glass, revealed a defect-free hybrid aligned liquid crystal cell on silicon substrate (Figure 4.12). An electrically tunable micro-resonator using photoaligned liquid crystal as cladding layers, where a photoalignment layer on the device surface defined the orientation of the liquid crystal molecules. The transmission property of the waveguide-coupled micro-resonator was electrically tuned by varying the cladding refractive index under an applied electric field in the vertical direction [1, 35].

The liquid crystal cladding refractive index is then varied according to the applied voltage, and subsequently the microresonator resonance wavelengths were tuned. Based on our initial measurements, the FSR (free spectral range) wavelength shift within the range of 20 nm was obtained, which is comparable with a thermo-optic effect (Figure 4.13). The new voltage controllable Si based add drop filters are envisaged based on this principle.

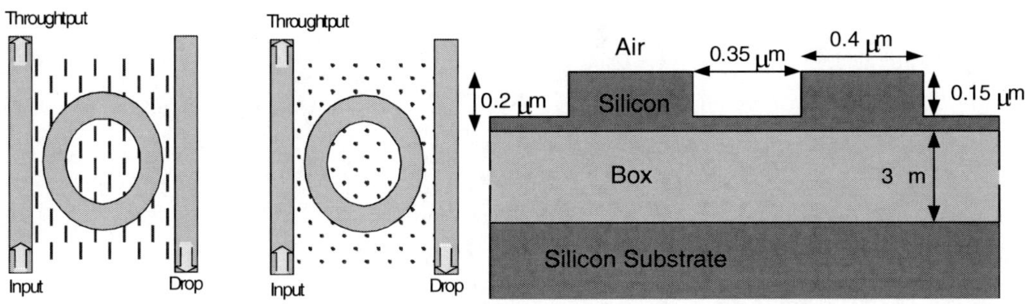

Figure 4.12. From left to right: Schematic profile of submicrometer-sized rib Si microring system; planar and homeotropic LC configurations on photoaligned cladding azo-dye layer [1, 35].

Remnant high efficiency polarization gratings were created in nematic liquid crystal cells by the layers of azo-dye molecules deposited on the cell substrates and exposed to "interfering" beams with opposite circular polarizations [39]. The diffraction picture was controlled by an electric field applied across the LC cell (Figure 4.14). The obtained polarization gratings can be used for electrically controlled discrimination and detection of polarized components of light. All molecules of LC were reoriented to a uniform homeotropic state at high voltage and there was no more modulation of LC alignment in the cell. Applications in LC optical switches are envisaged. The achromatic polarization grating (PG) exhibiting only three diffracted orders (0, ±1) can be also produced using photoalignment [40]. The achromatic PG offered the high efficiencies of thick (Bragg) gratings over a nearly entire range of a visible light. When used as an optical element of displays, the achromatic PG can be implemented as a switchable LC grating for modulator applications, by allowing one of the grating twists to be implemented as a switchable nematic LC and by placing the entire structure between the substrates with electrodes [40].

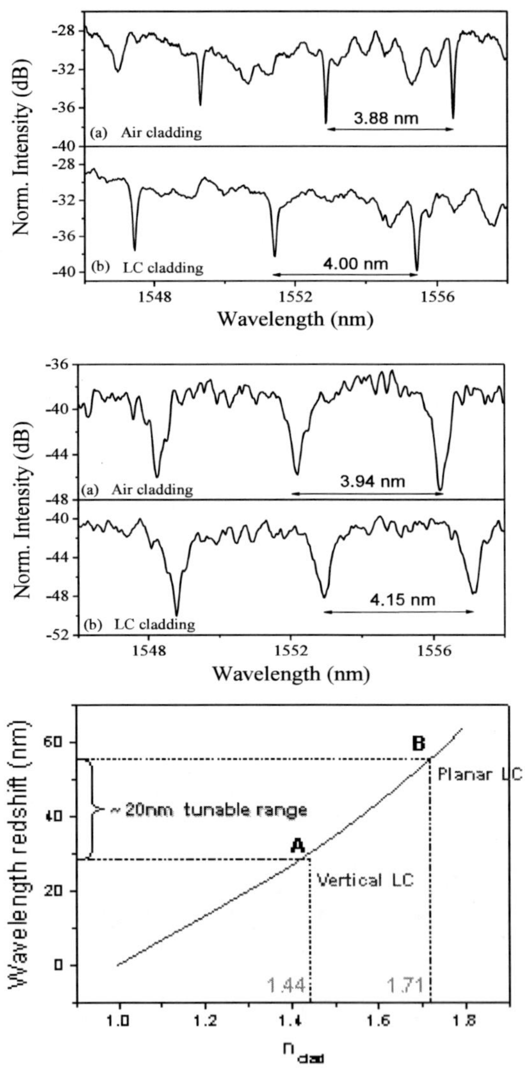

Figure 4.13. Characteristics of an electrically tunable micro-resonator using photoaligned liquid crystal as cladding layers. From left to right: FSR shift, when air cladding is changed to planar oriented LC; FSR shift, when air cladding is changed to homeotropic oriented LC; Wavelength redshift as a function of the effective refractive index of the LC layer [1,35].

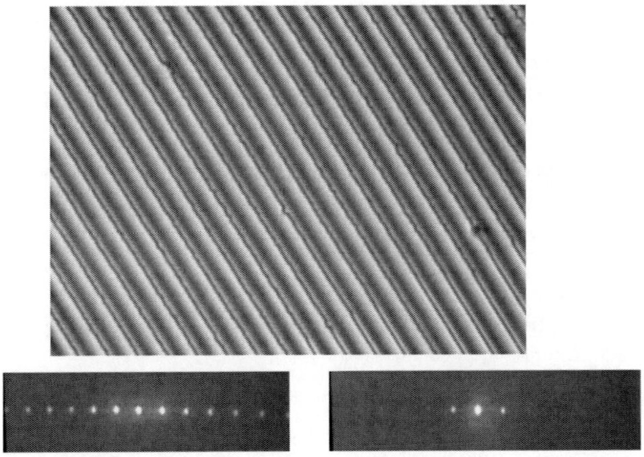

Figure 4.14. Tunable polarization gratings using liquid crystals [39].

Further work in the field of passive liquid crystal elements for fiber optical communication systems is needed. High-quality LC switches, variable optical attenuators, voltage controllable filters, based on Si Photonics devices, polarization controllers and rotators have to be developed. New working prototypes ready for packaging are highly desirable. Use of photonic crystals and photo-alignment technique makes it possible to develop new LC fiber components. There exist certain problems in wavelength dependence, too large losses, thermal drift, high crosstalk. New LC materials for the IR region used for fiber optical communications have to be tested. Specific LC materials with a high birefringence and low rotational viscosity have to be developed for IR region. The combination of photonic crystals and liquid crystals can provide effective switching of the refractive index of diffraction grating in the microsecond range with low controlling voltages, as the thickness of the LC layer in the hole can be less than one micron. LC aligning in the microtubes of photonic crystals can be produced by the photo-aligning technique. Fast PC/LC efficient switches and tunable filters are envisaged.

4.3. Liquid Crystal Lenses

Lenses with an electrically tunable focal length and/or are positionable focus are of broad interest for a number of applications. Because there are no moving parts in such a lens, they can be more reliable and able to withstand mechanical shock. Such tunable lenses can be applied in adaptive binocular glasses, autofocus cameras, LED lightsteering applications, 2D/3D switchable liquid crystal displays (LCDs) [41].

Many photonic devices required lenses with variable focal distance. A zoom lens system usually consists of a group of lenses, the separation distances between which are mechanically adjusted [42]. Mechanical adjusting processes are, as a rule, complicated and bulky. That is the reason for developing a new type of tunable lenses that are compact, lightweight, low-cost, and efficient. Such lenses are highly desirable and urgently needed for purposes of adaptive optics, optoelectronics, machine vision, stereo displays and eyeglasses applications. Due to the property to be controlled with an external electrical field LC cells are especially attractive for tunable lenses [43-46].

To obtain a switchable gradient profile within an LC layer, several methods have been proposed [44-46]. All of them can be classified into three categories: 1) lenses having a non-uniform profile of the cell gap and uniform applied voltage; 2) lenses with a planar LC cell and non-uniform spatial distribution of the applied voltage; and 3) lenses with non-uniform profile and non-uniform spatial distribution of the applied voltage. We will consider some of the lens configurations below.

4.3.1. Lens Based on Planar LC Cell with Non-Uniform Distribution of the Anchoring Energy

A tunable lens can be based on a planar uniform cell gap and is controlled with a uniform electrical field [43]. The needed gradient profile is achieved due to non-uniform surface conditions, in particular, spatial distribution of the anchoring energy. The LC lens was composed of LC material sandwiched between two transparent flat substrates. Similar to convention LC cells for displays, the inner sides of the substrates was covered with transparent electrodes and alignment layers. Specific of the LC lenses to be considered is that the anchoring energy of the alignment layers is not a constant over the surface and is characterized with some spatial distribution. The simplest case to achieve the distribution of the anchoring energy is the use of photoalightment technique [35].

The optical power D=1/f, where f is the lens focal distance of the lens having a large radius R can be evaluated with Fresnel's formula [20]:

$$D = \Delta\Phi\lambda/\pi R^2, \qquad (4.1)$$

where $\Delta\Phi$ is the phase difference between two extraordinary waves passed through the center and periphery of the lens, λ is the wavelength. The phase of an extraordinary wave is calculated as (see also (3.15)):

$$\Phi = \frac{2\pi}{\lambda}\int_L \frac{n_\perp n_\parallel dr}{\sqrt{n_\perp^2 \sin^2\theta(r) + n_\parallel^2 \cos^2\theta(r)}}, \ \Delta\Phi = \Phi - \Phi_0, \ \Phi_0 = 2\pi n_\perp/\lambda \qquad (4.2)$$

where L is the path along which the light ray propagates, $n_\perp$, $n_\parallel$ are the principal refractive indices of the liquid crystal, θ denotes the angle between the wave vector and the LC director. The LC lens is shown in Figure 4.15.

Eqs.4.1 and 4.2 show that it is possible to change the voltage V applied to LC cell thus changing the angle θ(r) distribution and phase retardation $\Delta\Phi$, which determine the lens focal distance (4.1). If the anchoring energy is non-uniform along the axis x parallel to LC cell surface (Figure 4.15), the lens profile will be created even for uniform driving voltage V over LC cell [43]. The optimal values of d and V to achieve a maximal possible optical power can be found for a strong anchoring energy (e.g. $3\cdot10^{-4}$J/m^2) and weak anchoring energy (e.g. $5\cdot10^{-6}$J/m^2) taking LC material E7 from Merck ($k_{11}=12\cdot10^{-12}$N; $k_{33}=19.5\cdot10^{-12}$N; $\varepsilon_\perp=5.1$; $\varepsilon_\parallel=19.6$; $n_\perp=1.5237$, $n_\parallel=1.75$) [43]. This material is good for LC lens, because of the high birefringence. The calculated phase difference $\Delta\Phi$, which is proportional to the lens optical power (4.1) as a function of the cell gap and applied voltage is plotted in Figure 4.16. The graph in Figure 4.16 shows that the phase difference $\Delta\Phi$ does not change critically versus the cell gap, when the last more then 10μm and versus the voltage in the range 0.8 - 2V. The phase difference $\Delta\Phi$ reaches the zone with the maximal value when the cell gap is around 15μm and voltage is around 1V. The further investigation was carried out for the LC cell with the cell gap d=15μm [43]. The value of the weak anchoring energy at $W=5\cdot10^{-6}J/m^2$ was fixed and the value of the strong anchoring energy varied. The dependence of the phase difference $\Delta\Phi$ versus applied voltage and the logarithm of the varied anchoring energy W is plotted in Figure 4.17. This dependence enables one to evaluate the needed distribution of the anchoring energy [43]. The smooth behaviour of the phase difference versus applied voltage makes possible to change the optical power of the LC lens in the wide range without essential distortions and defocusing.

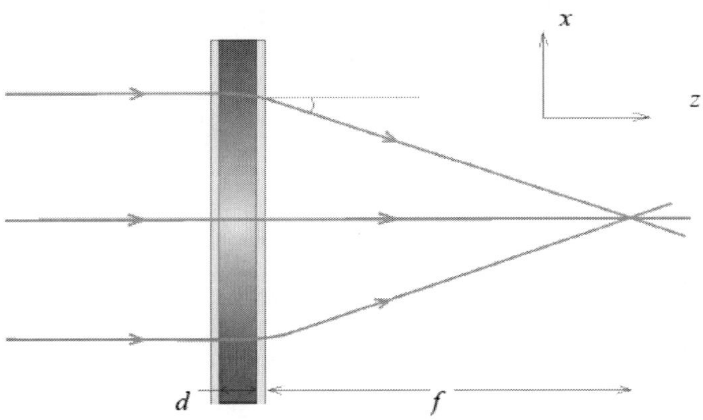

Figure 4.15. LC lens.

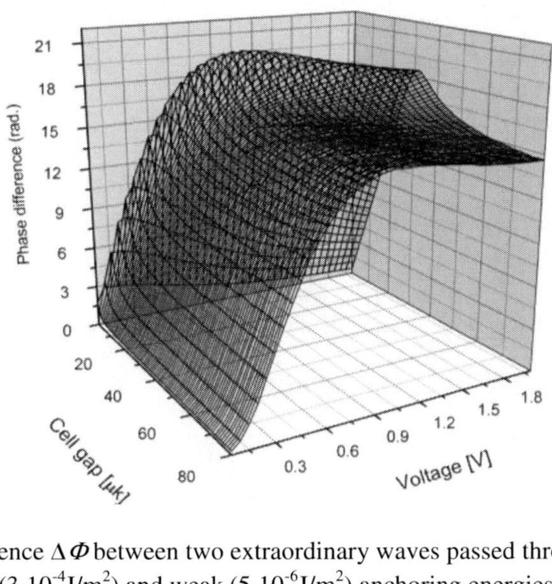

Figure 4.16. Phase difference $\Delta\Phi$ between two extraordinary waves passed through the LC cell in places having the strong ($3\cdot10^{-4}J/m^2$) and weak ($5\cdot10^{-6}J/m^2$) anchoring energies versus the cell gap and applied voltage [43].

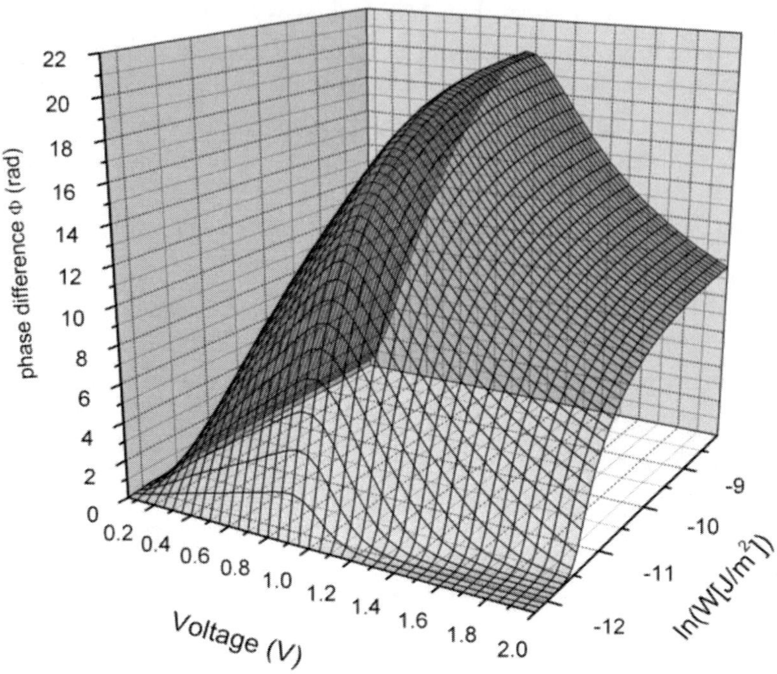

Figure 4.17. Phase difference $\Delta\Phi$ between two extraordinary waves passed through the LC cell in places with the weak anchoring energy ($5{\cdot}10^{-6}$J/m^2) and the varied one versus voltage and logarithm of the varied anchoring energy [43].

Using the graph presented in Figure 4.17, one can see that the phase difference $\Delta\Phi$ changes critically when logarithm of the anchoring energy varies from -12.2 to -10, that corresponds to W=$5{\cdot}10^{-6}$ J/m^2 and $5{\cdot}10^{-5}$ J/m^2, respectively. This new range is more tolerable for practical applications. Figure 4.18 contains results for both positive and negative lenses due to the nonuniform distribution of the anchoring energy [43].

A tunable LC lens can be also done using a non-uniform spatial distribution of the anchoring energy. The LC lens proposed has a uniform cell gap and voltage distribution. The method considered in Ref. [43] can be combined with any other known methods for reducing driving voltage or for simplifying the lens profile.

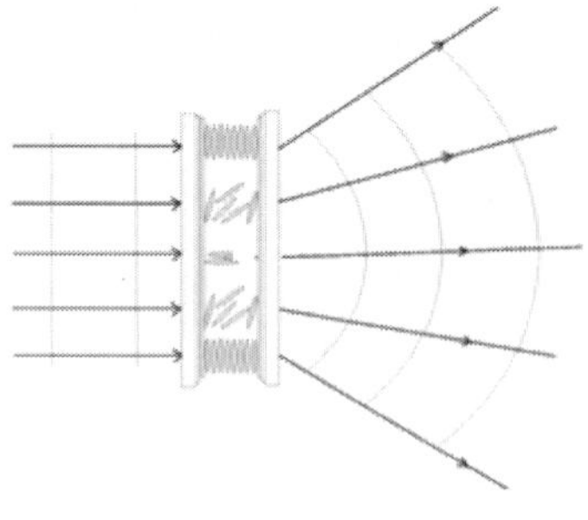

Figure 4.18. (Continued).

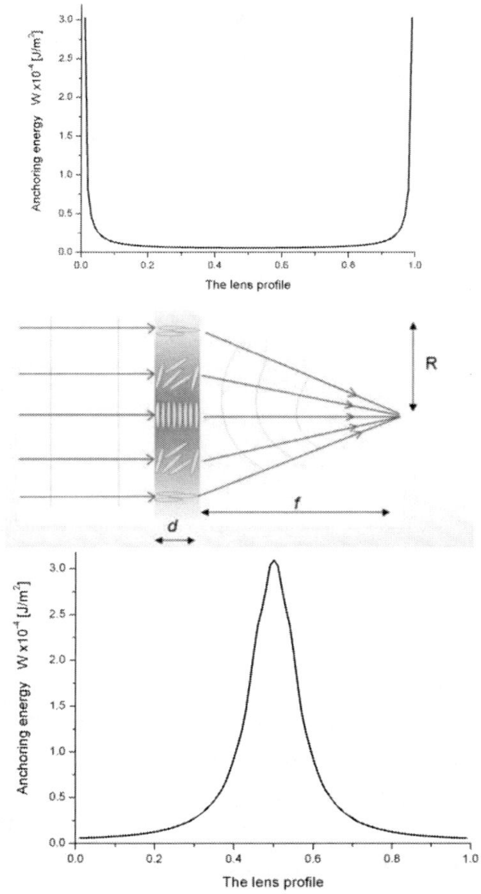

Figure 4.18. Tunable lenses (left) developed by a non-uniform distribution of LC anchoring energy (right [43]). Initial LC alignment is homogeneous. Top: diverging lens; bottom-converging lens.

4.3.2. Lens Based on Planar LC Cell with Non-Uniform Distribution of LC Alignment Angle

Stacked Alignment Layers

The stacked alignment layers consist of vertical and homogeneous alignment materials, can generate an intermediate pretilt angle [47]. Spatial variable UV dosage can be used to control the pretilt angle. An electrically tunable-focus liquid crystal lens is obtained using UV laser. A vertical alignment polyimide JALS2021 (JSR) was spin coated on an ITO glass substrate. The vertical alignment polyimide was a continuous layer since it completely covered the ITO crystalline pattern. The vertical alignment polyimide was then rubbed to obtain the first principle pretilt angle θ_0 (θ_0=85°). Then, a photo alignment material: ROP-103 (Rolic) was spin coated on top of the polyimide. The photo-alignment layer was exposed with different duration by a linearly polarized light with a wavelength of 340nm. Since some of the ROP-103 was not fully polymerized due to the short UV exposure time, the unpolymerized photo alignment material was rinsed away by the solvent. The longer the exposed time, the more photo alignment materials were polymerized and so that a thicker photo alignment was obtained. In addition, the sizes of domain which were formed by dewetting also varied accordingly with UV dosage. The resultant pretilt angle were varied with the UV dosage. Even with the same UV exposure time, if the UV light intensity is different, the pretilt angle will also be varied. The structure of the proposed spatially stacked alignment layers is shown in Figure 4.19.

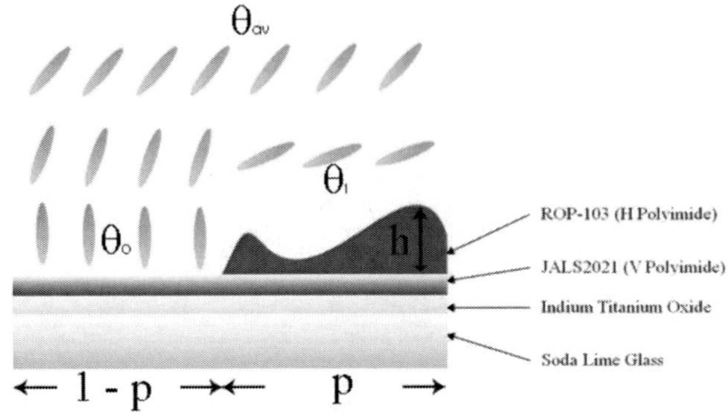

Figure 4.19. The double layer structure of the alignment film for lens applications [47].

It consists of two domains. One is a single continuous alignment layer and the other one is a top alignment layer which can be either continuous or discontinuous as controlled by dewetting. This type of alignment surfaces generally comprises two kinds of domains favoring different liquid crystal orientations. Suppose the area fraction of the top discontinuous layer is equal to p. As seen from Figure 4.20, the surface energy F_1 dominates when p is close to 1, so that the pretilt angle decreases when the stacked layer thickness increases.

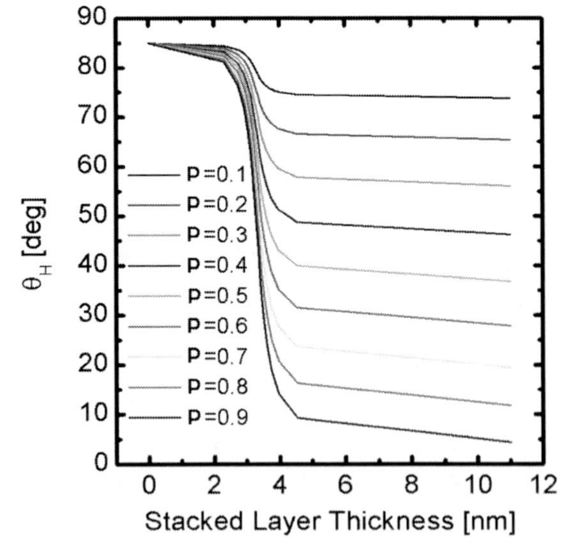

Figure 4.20. Pretilt angle variation as a function of a stacked layer thickness for different surface ratios p of photoalignment/polyimide film [47].

To show the lens effect, the LC lens was put between two polarizers which are perpendicular to each other. The rubbing direction of the bottom substrate was 45° with both of the polarizers. The light source with a green filter centered in 530nm was incident from one side of the polarizer and pass through the LC cell and the second polarizer. Microscope camera was used to take the photo of the LC cell. The transmittance information can be used to calculate the phase retardation of the LC lens. 1 kHz AC voltage source was added on the LC lens. Figure 4. 21 shows photos in different voltage.

The transmittance data was used to get the retardation profile, as shown in Figure 4.22. The x-axis parameter radius represents the distance from the center point. The y-axis parameter retardation is $\delta = \pi d\Delta n / \lambda$. The parabola approximation was made (pink curves in figure 4.22) for every retardation profile. We can see that in the radius range of 0.45mm, the retardation profile fits very well with the parabola curve in all the voltage, so in this area the lens is good, and 0.45mm radius circle was drawn in Figure 4.22.

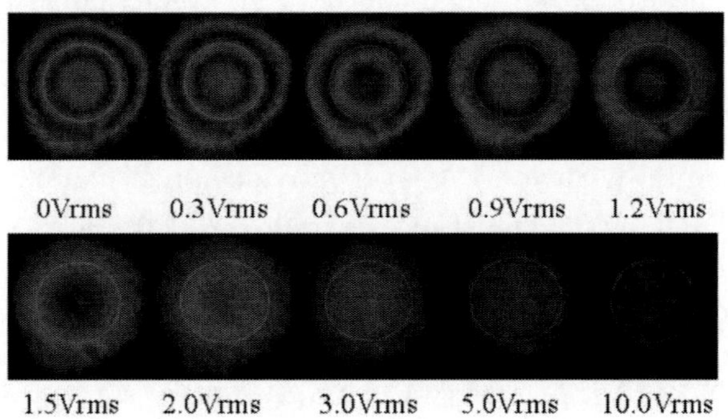

Figure 4.21. Transmittance of the lens structure between two polarizers for green light of 530nm [47].

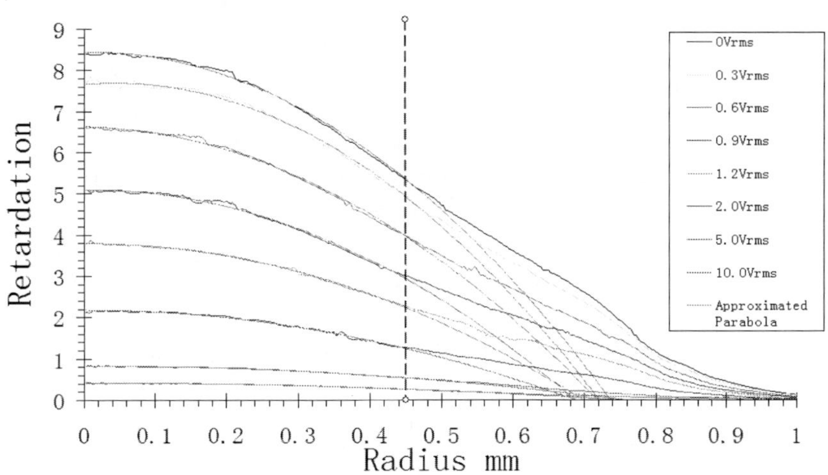

Figure 4.22. The retardation profile calculated from the intensity variation shown in Figure 3.3. The parabolic approximation is good for the size of the lens less than 0.45 mm [47].

The retardation profiles can be used to calculate the focal length for different voltages (4.1), as shown in Figure 4.23. The focal length was tunable from 17.2 cm to infinite [47].

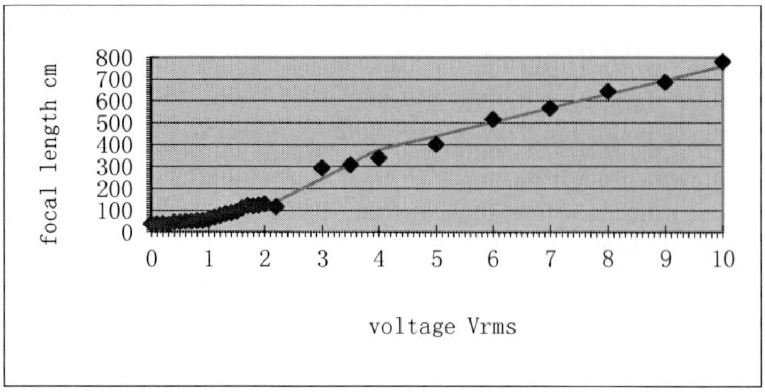

Figure 4.23. Tunable focal distance of the lens based on photoalignment materials [47].

Low Voltage Tunable Liquid Crystal Lens with a Single Alignment Layer

Spatially varying the LC pretilt angle can be a good alternative to achieve low power consuming tunable LC lenses [48]. The pretilt angle of the LC alignment is an important parameter to define the characteristic of LC devices. A tunable LC lens based on the approach of controlling the pretilt angle profile by means of the irradiance doses of a photoalignment (PA) layer was reported [48]. Such a PA layer provides precise control on the pretilt angle over a wide range of the pretilt angles, i.e., 1°–89°, which enables a great potential for applications not only in the tunable LC lens but also for other photonics and display devices. The PA material CPL024 (from DIC Corporation) was used, which is a cross-linking polymer, providing a broad range of pretilt angles (1°–89°) depending on the irradiance doses. The pretilt angle profile for the CPL024 at different irradiance has been studied by making a 10-μm thick-antiparallel LC cell with several distinct areas with different irradiance (Figure 4.24). The fabrication process includes the following steps. First, the CPL024 powder was dissolved in solvent cyclohexanone with a concentration of 0.5% wt/wt. Then, the solution was spin-coated on ITO glass substrates and thereafter exposed by UV light, of the wavelength 328 nm, at a plane inclined obliquely at an angle of 45°. This oblique exposure defines the azimuthal direction φ (defined in Figure 4.24) of the alignment layer which is critically important to generate such a regular pretilt angle profile; otherwise it will be random. The fabrication process flow is shown in the top insertion of Figure 4.24 [48].

Thereafter, the crystal rotating method was used to measure the pretilt angles for different areas with different irradiance; its dependence is shown in Figure 4.24. The pretilt angle increases with an increase in irradiance. The inset at the bottom right of Figure 4.24 represents the photos for the LC cell with different pretilt angles achieved at different irradiance, which confirms the good alignment quality with different pretilt angles [48]. The pretilt angle depends on the irradiance. Therefore, spatially variable UV irradiance patterns can be employed to achieve any desired pattern. The same approach has been used to fabricate the LC tunable lens (Figure 4.25) [48]. The schematics of the experimental setup and cell structure are shown in Figures 4.25(a) and 4.25(b), respectively.

To generate the spatially variable UV irradiance, a mask with circular structure of radius R was placed at a distance D from the substrate (Figure 4.25). The light intensity exposed on the substrate was distributed in circular symmetry with the smallest intensity at the center and the largest at the border of the circle. Therefore, after being exposed through this irradiance pattern, the pretilt angle defined on the substrate will be spatially variable with the smallest pretilt angle (1°) at the center and increasing gradually to the largest pretilt angle (89°) at the border. In the next step, the LC cell was assembled with the prepared substrates in an antiparallel azimuthal arrangement. The cell gap was maintained at 15 μm, and a nematic LC RDP97304 (from DIC Corporation) with $\Delta n = 0.27$ was filled into the LC cell. The fabricated tunable LC lens structure is shown in Figure 4.25. The microscopic photographs of the LC lens at different voltage are shown in Figure 4.26, wherein the length of the marker is 700 μm.

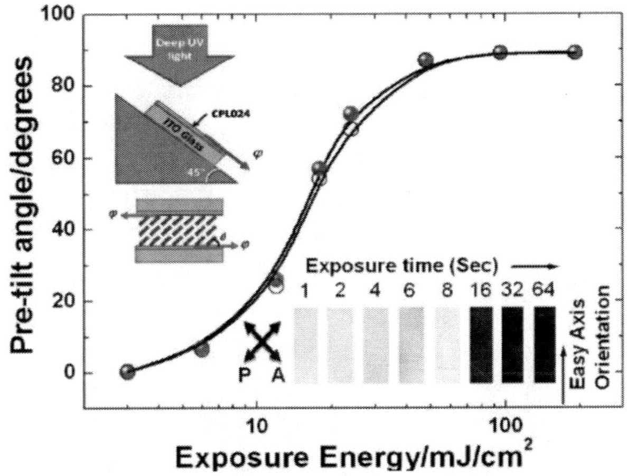

Figure 4.24. Pretilt angle dependence on the exposure energy (solid circles) [48]. Open legends represent the same experimental results repeated after several hours of photographing and thermal exposure. The top insertion shows the simple fabrication procedure to make the LC cell to analyze different PA parameters, whereas the bottom insertion shows the photograph of the prepared LC cell under the crossed polarizers for distinct areas with different irradiance and pretilt angles.

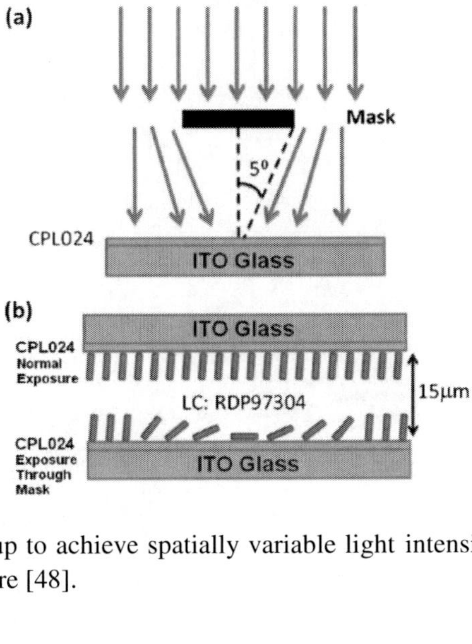

Figure 4.25. (a) Exposure setup to achieve spatially variable light intensity on the exposing plane. (b) The fabricated LC lens structure [48].

The retardation parameter profiles for the fabricated circular LC lens at different voltages are shown in Figure 4.27.

All of these retardation parameter profiles show parabolic character that is also the best case for any lens design. The solid lines in Figure 4.27 represent the best fit of the parabola approximation to the experimental retardation profile (open legends). It is clear from Figure 4.27 that the retardation parameter profiles fit very well with the parabola approximation, which confirms the working concept of the designed LC lens. The voltage dependence of the lens focal distance f is shown in the inset of Figure 4.27. Only 3 V is enough to tune f from 2.5 to 20 cm, which can be decreased further by optimizing the LC parameters and cell gap of the lens. Furthermore, given the easy fabrication procedure in association with low operating voltages, such devices could find applications in many modern optical and photonic devices. The proposed approach can also be employed to fabricate the bistable LC displays, photonic elements e.g., beam steering devices, diffraction gratings, etc. In spite of having several advantages, further research is required to improve the lens profile, which could include use of certain grayscale masks or control of exposing environments, but these are simple technological issues that can be resolved by further improvement in the technology.

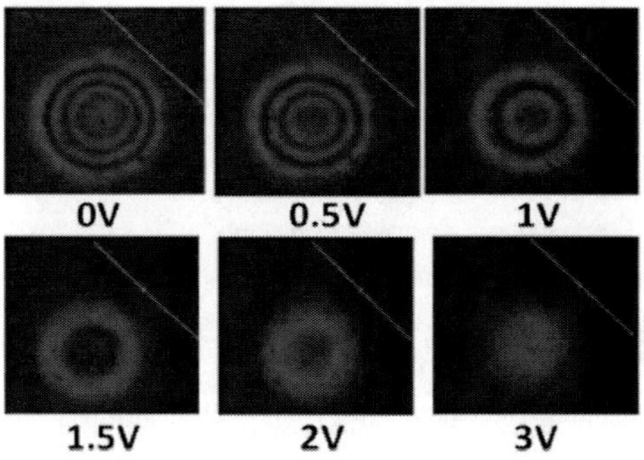

Figure 4.26. Optical microphotographs of the circular LC lens at different applied voltages that have been clicked in red light. The marker size is 700 μm [48].

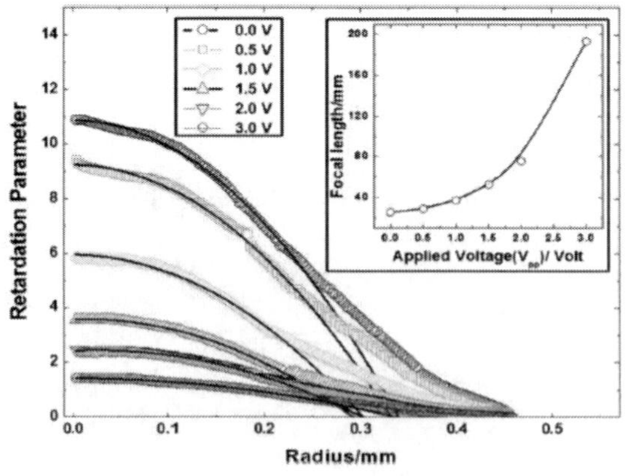

Figure 4.27. Retardation parameter profile dependence along the radius of the LC lens at different voltages. The open legends show the experimental retardation parameter profile, while the solid lines represent the best parabolic fitting. The inset shows the voltage dependence of the focal length of the same lens [48].

4.3.3. Switchable Fresnel Lens Based on Micro-Patterned Alignment

The photoalignment technology provides an opportunity for easy fabrication of the Fresnel zone lens structure, which is characterized by high efficiency and considerable fast response time. LC Fresnel zone lens with ~42% diffraction efficiency and ~6.7 ms response time was demonstrated [49]. The LC Fresnel zone lens was not only a spatial light modulator but also a polarization modulator. Furthermore, such Fresnel zone lens can be tuned electrically and optically as well. Such Fresnel lens was based on the alternatively twisted (Twisted Nematic mode) and planar (Electrically Controlled Birefringence mode) aligned domain (Figure 4.28). Figure 4.28 shows the configuration of liquid crystal in the cross section. The liquid crystal molecules in the white region are all in TN mode and those in black region are all in ECB mode.

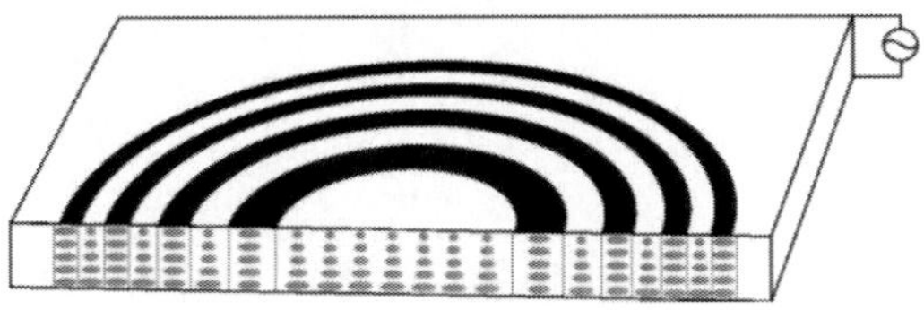

Figure 4.28. Configuration of the designed LC Fresnel zone lens [49].

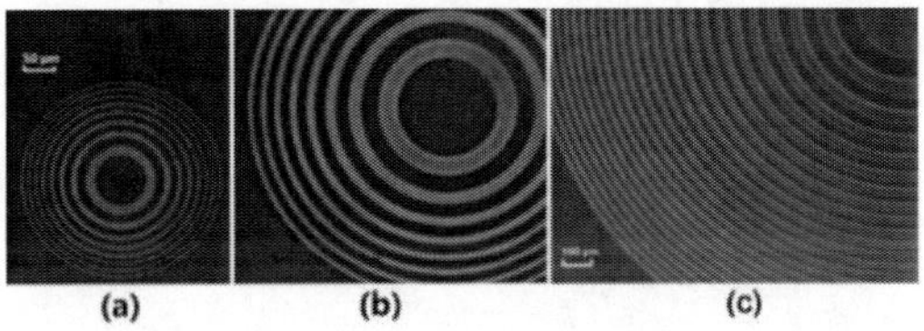

Figure 4.29. Optical microphotograph of the designed Fresnel zone lens under crossed polarizers [49]. (a) LC Fresnel zone lens (lens I) with the focal length of 3.2 mm, (b) the zoomed-in version of (a), (c) LC Fresnel zone lens (lens II) with the focal length of 50 mm.

Figures 4.29(a) and 4.29(b) represent the optical micrograph of the LC Fresnel zone lens (lens I) with the focal length of 3.2 mm under crossed polarizer with the easy axis of the planar aligned domain parallel to any of the polarizer axis. The black region represents planar aligned domain while the white circles represents the twisted nematic region. Figure 4.29(c) represents the optical micrograph of the LC Fresnel zone lens (lens II) with the focal length of 50 mm. Figure 4.30 shows the diffraction pattern of the proposed structure in diffractive 4.30(a), 4.30(f), and non-diffractive states 4.30(b), 4.30(g).

The on-time of the Fresnel LC cell was about 100 μs and the off-time is about 6.6 ms, so the total response time is ~6.7 ms which is relatively fast as compared with the existing devices [49]. The diffraction efficiency was ~42%, which could be further improved and get approached to the theoretical diffraction efficiency ~60%.

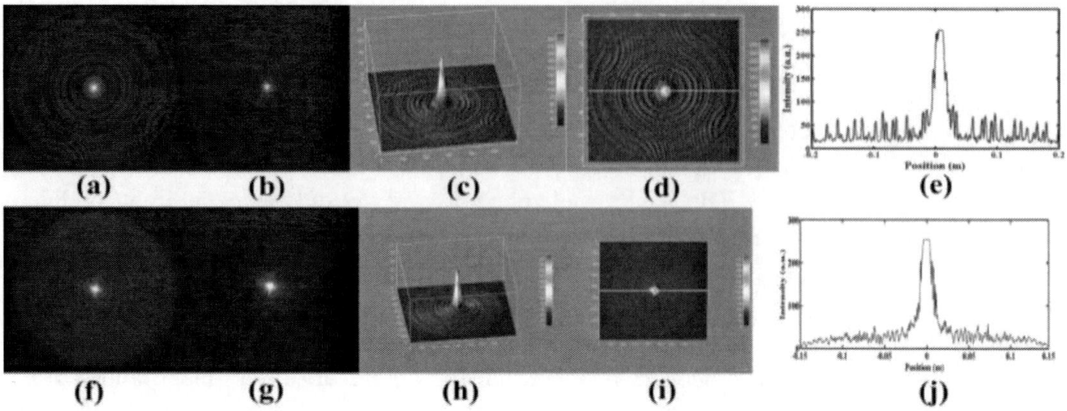

Figure 4.30. Diffraction pattern for the designed LC Fresnel zone lens [49]. (a) diffractive state of lens I, (b) dark or non-diffractive state of lens I at the electric field of 10V, (c) 3D light intensity profile for lens I, (d) 2D top view of (c), (e) light intensity distribution along white line in (d), (f) diffractive state of lens II, (g) dark or non-diffractive state of lens II at the electric field of 10V, (h) 3D light intensity profile for lens II, (i) 2D top view of (h), (j) light intensity distribution along white line in (i).

4.3.4. Polarizer Free Lens Liquid Crystal Display

In a classic lens design, the optical path difference across the lens aperture has a profile, such as sphere or parabola, that's determined only by the relief of the material surface, and the index of refraction is homogeneously constant [50]. The optical power D that can be achieved in a liquid crystal lens of the radius R can be evaluated with the formula (see also (4.1):

$$D = \frac{1}{f} = \frac{2d(n_e - n_o)}{R^2}, \tag{4.3}$$

where f is the focal distance of the lens, d is the thickness of the layer of the liquid crystal material, and n_e and n_o are the principal indexes of refraction of the liquid crystal material.

By achieving a specific distribution of the pretilt angle (see also 4.2), it is possible to obtain the needed distribution of the molecules orientations that gives the lens effect. When a high voltage (more than the saturation voltage) is applied, all molecules of liquid crystal material will be oriented perpendicular to the substrates (if the liquid crystal possesses positive dielectric anisotropy) or parallel to the substrates (if liquid crystal possesses the negative dielectric anisotropy). In this case, the layer of liquid crystal acts as a transparent plane-parallel plate and has no effect on the direction of the light propagation.

The same non-uniform distribution can be done e.g. by placing the dielectric layer with a non-uniform thickness on one of the layer in order to create a nonuniform electric field (Figure 4.31).

Figure 4.32 is a schematic sectional view of a pixel as a polarization independent binary Fresnel's lens (see also Figure 4.28) [50]. This lens consists of domains, which correspond to Fresnel's zones and depend on position in the lens. Orientations of molecules are orthogonal in two adjacent zones.

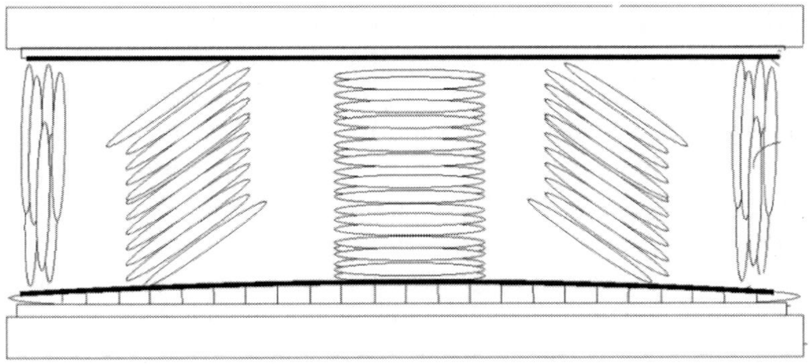

Figure 4.31. Non-uniform distribution can be done e.g. by placing the dielectric layer with a non-uniform thickness on one of the layer in order to create a nonuniform electric field [50].

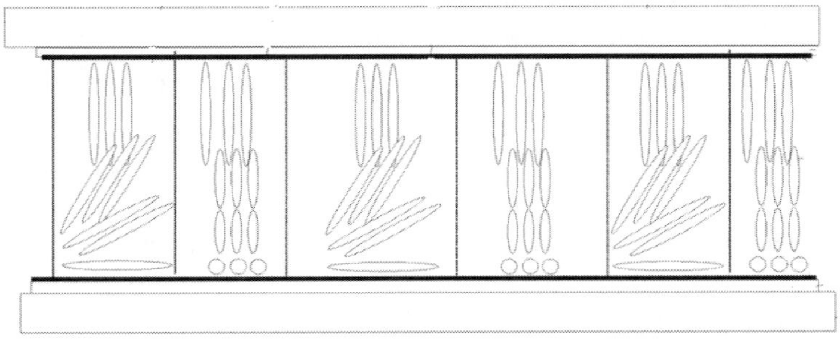

Figure 4.32. A polarization independent binary Fresnel's lens [50].

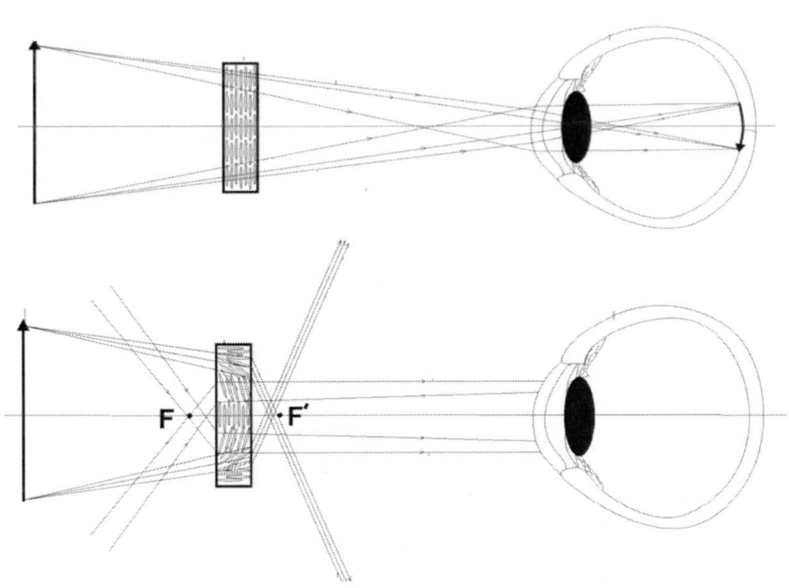

Figure 4.33. Up: A schematic sectional view of the rays tracing from an object that is seen through a display when optical power of pixels is zero and formation of the image of this object on the retina; Down: A schematic sectional view of the rays tracing from an object that is not seen through a pixel when optical power of pixel is big and positive: the pixel acts as convergent lens. The same effect is observed, when pixel is large and negative: the pixel acts as divergent lens [50].

The principle of the polarizer free lens liquid crystal display is shown in Figure 4.33 [50]. If a pixel is not activated, LCD cell is transparent as shown in Figure 4.33 (up) . If the pixel is activated, it is not transparent as shown and form on the eye retina a grey spot in the place determined by the geometrical position of the pixel with respect to the eye (Figure 4.33). The lack of the image on the pixel can be done by creating either divergent or convergent lens in accordance with the ray tracing picture shown on Figure 4.33. Figure 4.33 is a schematic sectional view of the ray tracing from an object located outside in the case when the pixel acts as a short focus convergent lens. In this case, the rays coming from the object pass the pixel and do not reach the eye. The rays coming from outside reaches the eye. The eye sees the pixel as a spot, brightness of which is equal to average brightness of surrounding environment (Figure 4.33).

Figure 4.34 is a schematic sectional view of the lens non-polarizer display with background (lion picture). The brightness of the image (display image) is equal to the average brightness of the background (lion picture).

Thus a lens non-polarizer liquid crystal display containing an array of pixels was proposed [50]. Each pixel contains at least one layer of liquid crystal with two electrodes; alignment layers setting orientations of the molecules of said liquid crystal near its boundaries and covering substrates, wherein the pixels are lenses with controlled optical power. A lens can be created by a non-uniform distribution of the LC anchoring energy, non-uniform electric field or non-uniform thickness of the LC or alignment layer. Double lens construction is also possible. Photoalignment technology can give a unique opportunity to get a non-uniform LC alignment either by non-uniform exposure or non-uniform thickness of the alignment layer (see also 4.2). The polarizers are not necessary for this kind of LCD. Random azimuthal distribution of the LC molecules is also possible.

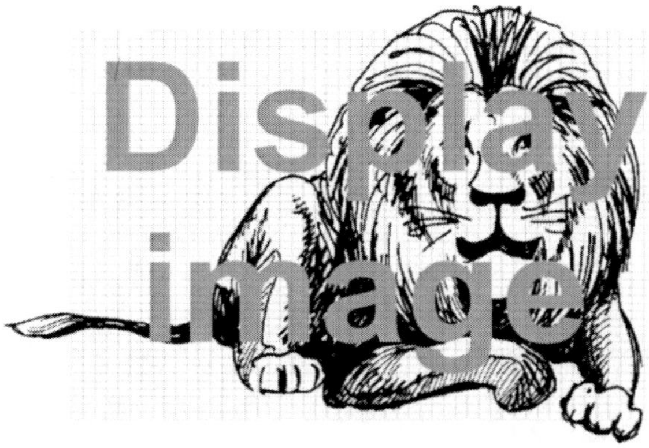

Figure 4.34. A schematic view of the image seen by observer [50].

4.3.5. Other LC Lenses

Lenses with an electrically tunable focal length and/or a repositionable focus are of broad interest for a number of applications. Because there are no moving parts in such a lens, they can be more reliable and able to withstand mechanical shock. Such tunable lenses can be applied in adaptive binocular glasses, autofocus cameras, LED lightsteering applications, 3D confocal microscopy, 2D/3D switchable LCD etc [51]. A wide number of approaches have been studied to realize tunable lenses, and some of them are illustrated below.

Lens with a Special Form of Electrodes

Figure 4.35 [51] shows a configuration with a LC layer with nonuniform thickness (see also Figure 4.31). The biggest problem with LC is the fact that such lenses only work with one linear polarization component of the light so that a polarizer is necessary.

Figure 4.35 (b) shows a different approach in which the LC layer is homogeneous. Because of curvature of the glass and, hence, the nonuniform distance between the electrodes at a certain position in the cell, the LC switches inhomogeneously and this results in a variable focal length. Figure 4.35(c) shows a different approach that is nowadays succesfully used in autofocus cameras in cell phones [51]. The LC layer is again homogeneous, but due to the use of two different dielectric materials with different dielectric constants, the electric field in the LC layer is different at different positions. The refractive index of the dielectric materials are matched, so the LC layer itself fully incorporates the optical functionality, while the two dielectric materials take care of the electric functionality.

The other possibility to produce the lens effect is to create a non-uniform electric field using patterned electrodes. Fast response superzone Fresnel LC lens with multiple transparent electrodes was proposed for temporal scanning auto-stereoscopic display [52]. The experimental results indicated that the superzone Fresnel LC lens not only performed fast switching time (0.2 s), but also had the benefit of low driving voltage about 5V. A 4-inch 2D/3D switchable auto-stereoscopic display with superzone Fresnel LC lens was further demonstrated [52]. Finally, by driving the multiple electrodes alternatively, the superzone Fresnel LC lens array could be moved on horizontal direction for increasing the resolution of auto-stereoscopic display. Figure 4.36 shows, how changing the electrode configuration, it is possible to optimize the phase profile of the Fresnel lens.

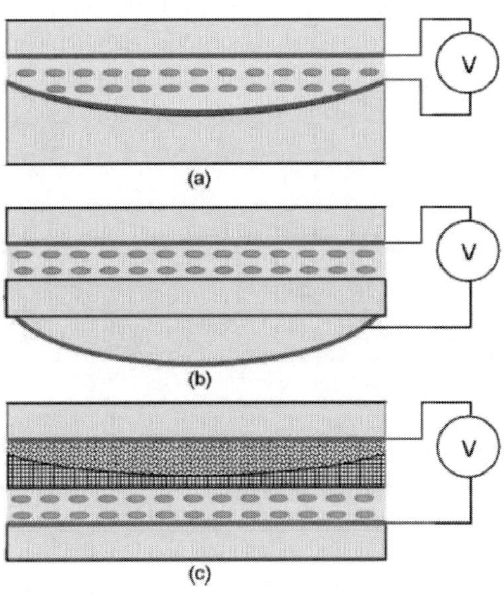

Figure 4.35. Different approaches for tunable lenses with liquid-crystals [51].

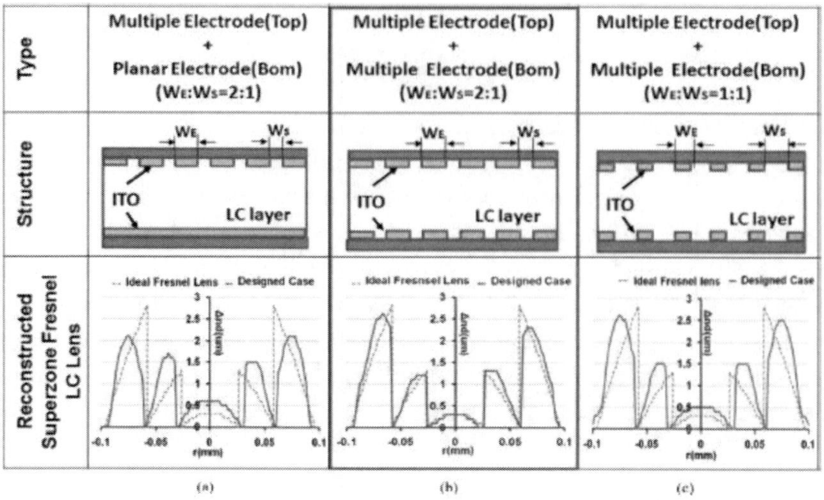

Figure 4.36. Simulated cases for finding out the optimized structure design—Multiple electrode on both top and bottom side of substrate. The optimized phase profile is shown below [52].

A high-resistance liquid crystal (HR-LC) lens can be is utilized to obtain a parabolic profile of LC lens [53]. A high-resistance material (e.g. IGZO semiconductor material) was coated on the controlled electrodes layer, as shown in Figure 4.37. Using the high-resistance layer, the gradient changed potential distribution can be easily produced when applying the different voltages between the center electrode, planar electrode, and edge electrode (see Figure 4.37). Thus, the LC molecules will reorient according to the electric field. Finally, the parabolic curve of the lens phase profile can be approached. HR-LC lens whose diameter and LC cell gap was designed as 2 mm and 60 μm, respectively provides the focal length variation between 4 cm to ∞, when the switching driving voltages from 5 V to 0 V was applied [53].

Several more lenses with various form of electrode configurations such as : double LC layer with perpendicular LC orientations to enable the lens effect for the arbitrary light polarization [54]; LC lens with spherical electrodes (see also Figure 4.35) [55] ; microlens array lens with fast focal switching using ferroelectric LC [56]; LC lens with curved conducting polymer electrodes [57] were also developed recently.

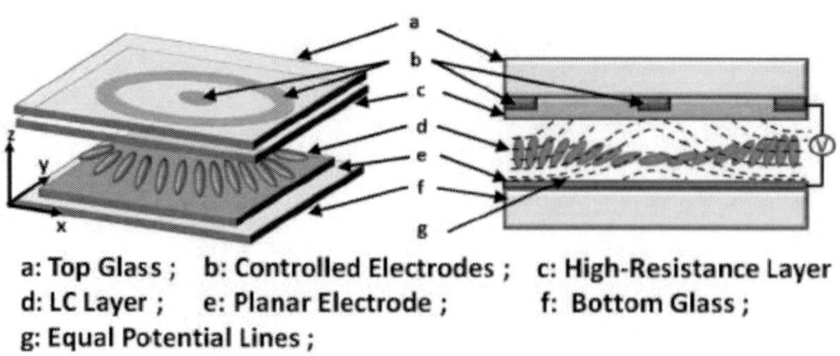

Figure 4.37. Structure of high-resistance liquid crystal (HR-LC) lens. High-resistance layer is coated on the controlled electrode layer [53].

Lens with a Polymer Concentration Gradient Inside LC Layer

Another lens configuration is a uniform LC layer with uniform electrodes, and the nonuniformity of the LC switching is obtained by a gradient in the polymer concentration of a polymer stabilized liquid-crystal [58]. The lenslike gradient refractive index profile is created in the cell via the spatially distributed polymer network obtained by photopolymerization using a spatially nonuniform laser beam. The conditions of the polymer network generation was optimized to improve the optical quality of the lens and its focusing properties. Low optical loss scattering is achieved for a focal length smoothly tunable from infinity to 0.8 m. Obtained results can be applied to develop lenses that have no moving parts and allow the electro-optical zooming. The example of such lens is shown in Figure 4.38 [58].

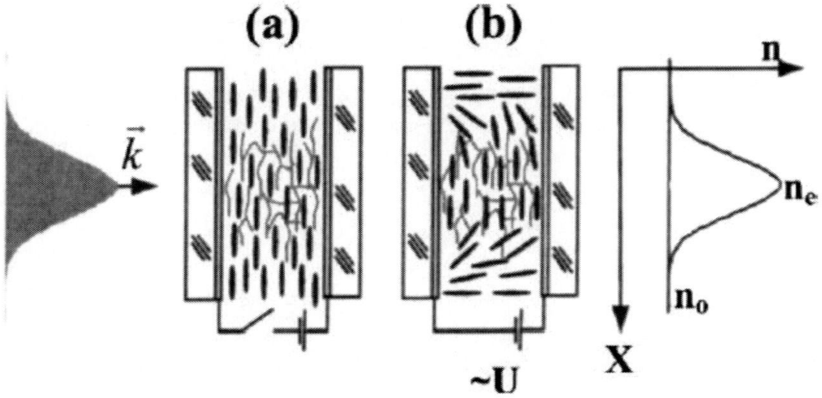

Figure 4.38. Schematic demonstration of the principle of (a) the PSLC lens fabrication and (b) its driving with a voltage and resulting refractive index distribution [58].

The focal length of such a lens may be efficiently controlled from infinity down to 80 cm by the amplitude and the frequency of the driving ac voltage. Two drawbacks remain to be solved before considering wide range applications of such lenses, i.e. polarization dependence and residual scattering [58].

Tunable electronic lenses using gradient polymer network liquid crystal (PNLC) cells were demonstrated [59] . By changing the photomask pattern, both positive and negative lenses were fabricated (Figure 4.39). The optical density (OD) was decreased outward radically from a maximum OD=1.0 at the center to OD=0.04 at the edges. The output intensity was varied in a parabolic profile across the sample. The area with a stronger UV light would accelerate the polymerization process and produce a higher polymer concentration. Conversely, the area with a weaker UV exposure would have a lower polymer network concentration. As a result, a polymer network with inhomogeneous centrosymmetric morphology was formed.

The advantages of such a PNLC lens were low operation voltage, large aperture size, and simple electrode design. To overcome the polarization dependence, the stacking two orthogonal homogeneous PNLC cells were also considered [59].

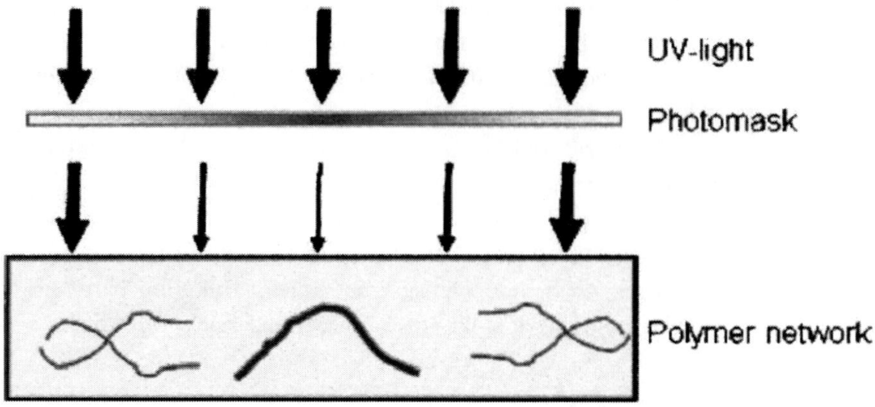

Figure 4.39. Fabrication of an inhomogeneous PNLC using a centrosymmetric circular continuously variable density filter as photomask. OD=1 at center and 0.04 at edges [59].

4.4. LC Sensors

Liquid crystals (LC) can be considered as very perspective materials for sensor applications [14]. It is determined by high sensitivity of liquid crystal media to the action of different factors of physical (or chemical) nature. In general, a sensor is often defined as a device that receives and responds to a signal or stimulus [59]. Stimulus is the quantity, property or condition that is sensed and converted into definite (mostly electrical) signal. Sensing and visualization of temperature fields via cholesteric liquid crystals can be considered as the most traditional sensor application of liquid crystals. It is based on a strong sensitivity of the selective reflection spectra to temperature variations which makes possible to visualize temperature fields [13,14]. Such possibility is traditionally used in biomedical applications of liquid crystals [60] for a diagnostics of different kinds of pathologies. It was also applied for registration, visualization and measurements of air flows which modify the temperature fields [14]. Cholesterics and nematic LC were also proposed for the fiber optic pressure sensors operating at high pressures as well as chemical and biological sensors which presently attract a growing interest [14]. We will consider in this Chapter several new recent LC configurations for sensing applications.

4.4.1. Sensors at the Fibre Tips: A Novel Liquid Crystal-Based Photonic Transducer for Sensing Systems

Based on recent advancements in nano-materials, nanofabrication and new methods of liquid crystal photo-alignment, a fully integrated polarisation-based photonic transducer, which tends itself to multiplexing and scalable fabrication was proposed [61]. Its implementation takes the form of a sensor head, depicted in Figure 4.40, which uses a DHF liquid crystal (see also 3.5.2.4) plate coupled to a standard single mode optical fibre via a nano-wire grid polariser (NWGP) where a novel non-contact photo-alignment technique allows a precise liquid crystal alignment with respect to the nanostructure (see also 4.3.5).

This voltage-sensing reflective cell offers a continuous hysteresis-free optical phase shift and remarkable linearity of the reflectance response. The cell can also operate using unpolarised broadband illumination and consequently can be multiplexed using standard single-mode fibres, avoiding the need for polarisation maintaining fibres. Moreover, the sensor operates at low driving voltages, i.e. from 0 to 3 V, enabling the device for a wide range of applications.

The reflected light intensity shows a symmetrical V-shaped response if the angle between the polarizer and FLC helix axis $\beta = 0$ (see Fig. 3.40), while a linear dependence of $R_n(V)$ manifests clearly within ±3 V when β is around 15°–17° (Figure 4.41). Our estimations show, that the surface alignment strongly influences the electro-optical properties of DHF FLC cell.

The following optical sensor head (Figure 4.42) for novel sensing technologies including voltage sensing (smart grid), hydrophones, sonars and pressure sensors can be made on the basis of DHF FLC reflecting cell.

In the proposed configuration, the liquid crystal cell reflectance is modulated by the voltage applied to the electrodes and it can be easily measured by means of the integrated single mode fibre using broadband unpolarised light at the standard telecommunication wavelength λ=1.55 μ (Figure 4.43). It is easy to show that the response (in our case, reflectance) of the cell to the applied electric field is in general non-linear, but operating conditions can be chosen to simultaneously maximize the reflectance modulation and obtain a linear response, which is crucial when operating the LC cell as a sensor. This is true in the so-called weak field limit, when the applied electric field E is small with respect to the critical field of the helix unwinding (see also (3.63)).

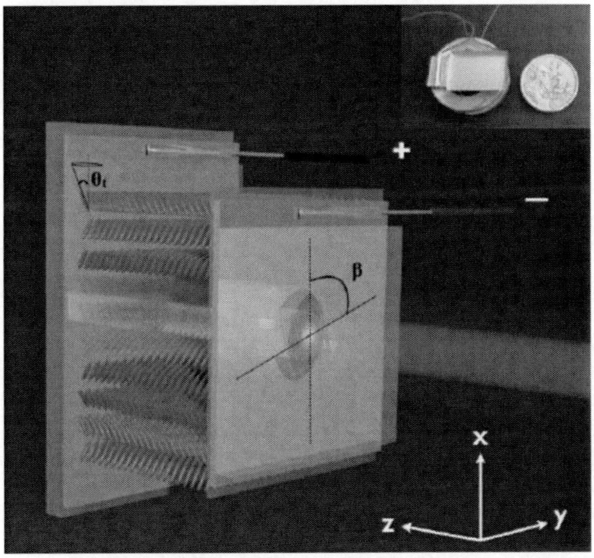

Figure 4.40. Device architecture [61]. The device is a layered structure comprised between two glass substrates. An optical fibre is glued to the front glass substrate to inject light through the following layers: the front gold electrode (–), on which a nano-wire grid polariser is etched, a photo-alignment layer (light blue), a liquid crystal plate of thickness 5 μm, a second alignment layer (light blue), the back gold electrode (+) which also acts as a broad band mirror and reflects light back into the fibre. The liquid crystal is represented as a stack of smectic layers made of ellipsoidal molecules (not to scale). In the Figure, the transmitted polarisation forms an angle β with the LC helix axis, which is along x (Fig. 3.40). The DHF FLC tilt angle, θ_t is also shown. Inset: a picture of the device.

The application of an electric field perpendicular to the cell plane essentially amounts to a change in the electrically controlled effective birefringence, Δn, and to a rotation of the cell optical axes by an angle α around an axis parallel to the field. The reflectance R at parallel analyser/polariser can then be expressed as (see also (3.66)):

$$R = 1 - \sin^2(2\pi\Delta n(z)d/\lambda)\sin^2(2(\beta-\alpha(z))) \quad (4.4)$$

where β is the angle between the polariser's direction and the undistorted helix axis (see also Figure 3.40).

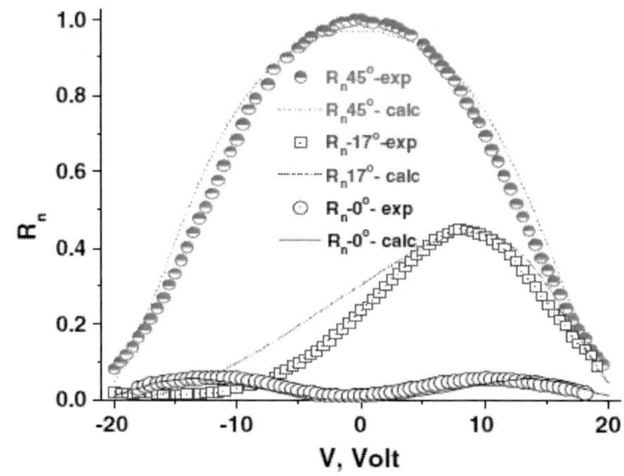

Figure. 4.41. Measured (symbols) and calculated (lines) dependencies of the normalized reflection on the square voltage amplitude applied to 5 μm thick FLC-576 layer injected into the reflected DHFLC cell. Measurements have been carried out at frequency 500 Hz and $\beta = 0°$, 17°, and 45° [61].

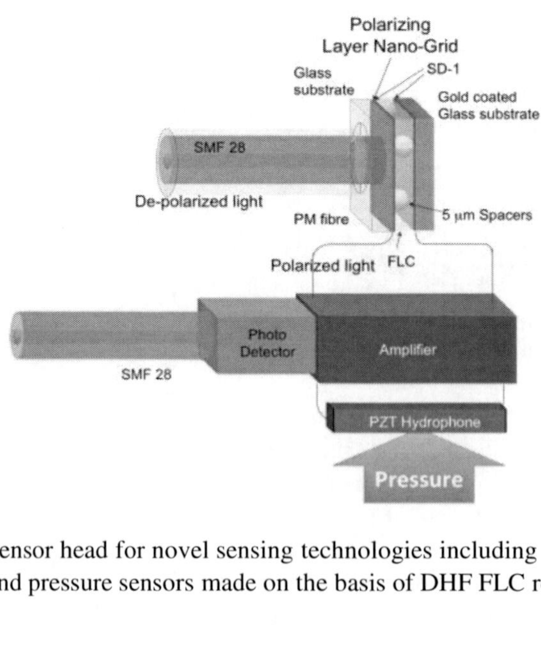

Figure. 4.42. Optical sensor head for novel sensing technologies including voltage sensing (smart grid), hydrophones, sonars and pressure sensors made on the basis of DHF FLC reflecting cell [61].

The cell reflectance at λ= 1.55μ as a function of the static voltage V_0 applied to the cell electrodes was measured [61]. The results are reported in Figure 4.43 for the applied voltages up to 10 V. It was seen, that for voltages $|V_0|$ < 0.5 V, the cell exhibits an excellent linear response and that the numerical results are in good agreement with the experiment, i.e. the proposed device has an extended linear region around V=0, no bias voltage is required [61]. The simulated reflectance at β = =-13^0 is also shown in Figure 4.43 and it can be considered as the most linear response achievable for a 5 μm cell thickness. Both linearity and modulation amplitude can be further improved by optimizing the sensor head design.

For many applications, it is important that the sensor has a fast response time, allowing it to work at high frequencies. The optical response at various frequencies was applied with a sinusoidal driving voltage V (t) = $V_0 \sin(2\pi\nu_0 t)$ of frequency ν_0 and the power spectrum of the response signal, defined as the magnitude of its Fourier transform was measured (Figure 4.44) [61]. The signal shows an excellent linear behaviour up to V_0 = 3 V, due to the fact that the quadratic term of the cell response to the voltage does not contribute to the first harmonic and the dynamic range of the present device is much larger than in the static case (Figure 4.44).

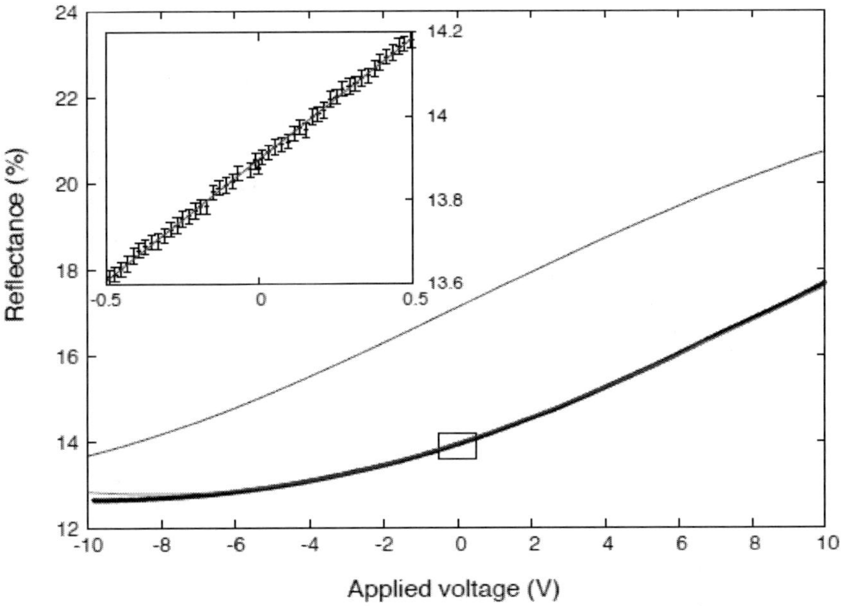

Figure 4.43. Reflectance measured (black solid symbols) and simulated (solid lines) as a function of the voltage applied to the cell. The red and blue solid lines have been calculated for β =-18^0 and β = =-13^0, respectively. Inset: zoom of the small voltage region, where the cell exhibits an excellent linear response. The solid line is the best linear fit for this region. Error bars represent one standard deviation [61].

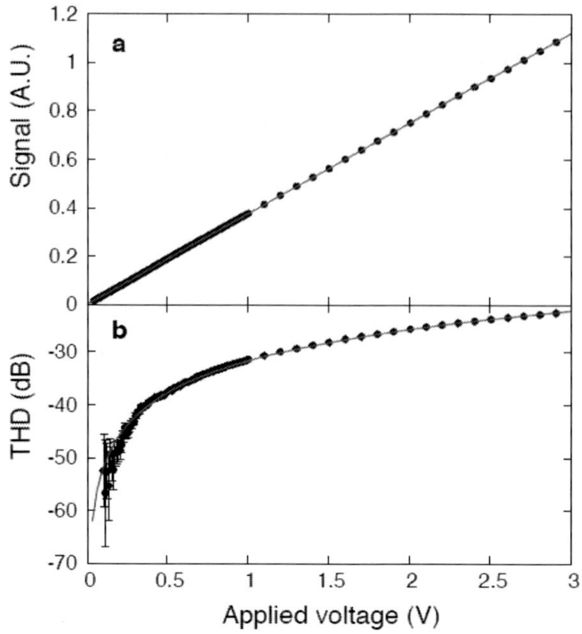

Figure 4.44. (a) Amplitude of the first harmonic of the response signal measured, for a driving voltage of frequency ν_0 = 1 kHz, as a function of the driving voltage amplitude (black circles). The red line is the best linear fit. (b) Measured Total Harmonic Distortion at 1kHz as a function of the driving voltage amplitude (black circles). The red line represents the best fit [61].

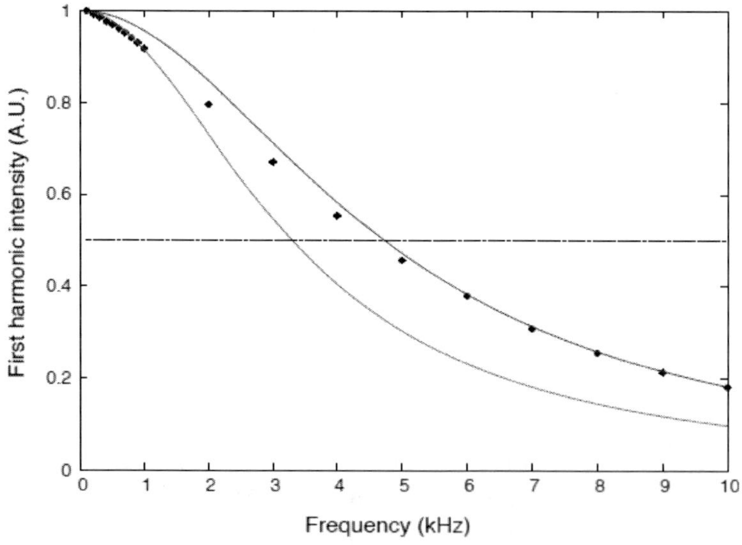

Figure 4.45. Intensity of the first harmonic signal as a function of the driving voltage frequency. The dashed line indicates an intensity corresponding to half the static value and it is a guide for the eye [61].

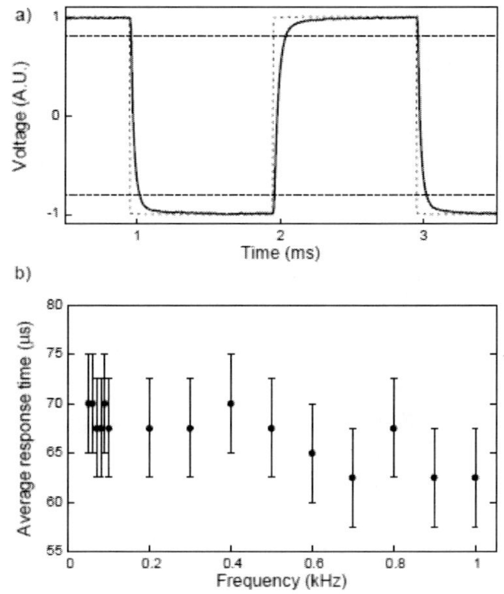

Figure 4.46. a) Typical response time measurement showing the applied square wave voltage (dashed line) and device response (solid line), both normalized. Dotted lines indicate 10% and 90% of the step height. b) Average response time of the device as a function of the applied voltage frequency [61].

Figure 4.45, shows that the signal intensity decreases at high frequencies. The device bandwidth, defined as the frequency at which the signal intensity becomes ½ of the static value, is about 4.5 kHz (Figure 4.45) [61].

The average response times for frequencies between 50 Hz and 1 kHz have been found to range from 70 μs to 60μs and shown in Figure 4.46 [61].

Thus using nano-grid electrodes, nanometer-scale SD1 photo-alignment layers and DHF FLC with a sub-wavelength helix pitch of 150 nm help to realize a fully integrated polarization-based photonic transducer which exhibits fast response to an applied voltage/electric field, excellent linearity and allows multiplexing using standard, low cost telecom optical fibers. Continuous and hysteresis-free optical phase shift with response time potentially as low as 1 μs is possible. A high voltage monitoring or,when combined with a piezo-actuator, flow measurement, hydrophones and towed array sonar systems becomes possible.

4.4.2. Liquid-Crystal Micropolarimeter Array for Full Stokes Polarization Imaging in Visible Spectrum Based on Photopatterned Azo-Dye

A liquid-crystal based micropolarimeter (LCMP) array for full Stokes polarization imaging in visible spectrum was demonstrated [62,63]. The proposed micropolarimeter was fabricated by patterning a liquid-crystal (LC) layer on top of a visible-regime metal-wire-grid polarizer (MWGP) using photoaligned sulfonic azo-dye SD1 [35] as the LC photoalignment material. This arrangement enables the formation of either micrometer-scale LC polarization rotators, neutral density filters or quarter wavelength retarders. These elements are in turn exploited to acquire all components of the Stokes vector, which describes all possible polarization states of light. The proposed LC micropolarimeter array can be integrated on top of a complementary metal-oxide-semiconductor (CMOS) image sensor for real-time full Stokes polarization imaging [62,63].

Figure 4.47 illustrates how the proposed LCMP array can be integrated on top of a CMOS image sensor to extract the full Stokes components for every pixel of the captured image.

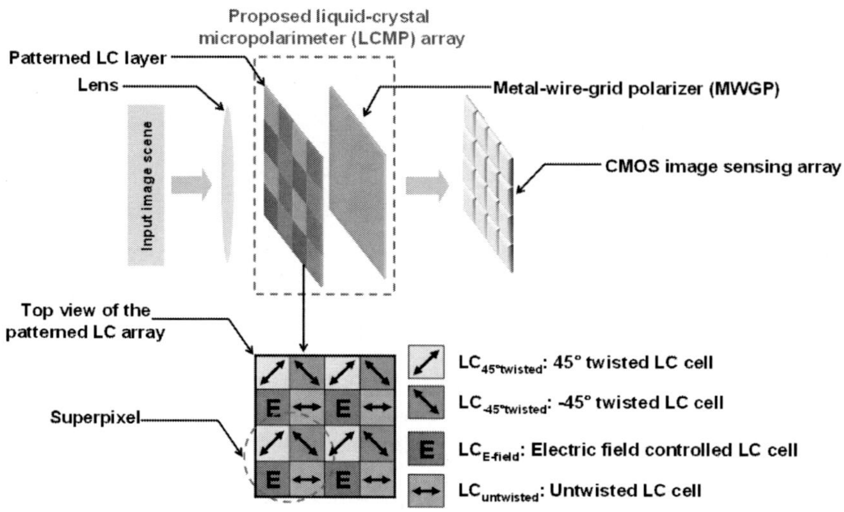

Figure 4.47. Photo-patterned micropolarizer array for a simultaneous detection of all four Stokes parameters of output optical image [62,63].

Figure 4.48 shows cross-sections for the four micropolarimeters. Here 45° twisted, −45° twisted and untwisted LC cells were chosen for three out of the four micropolarimeters covering a superpixel. The fourth LCMP operated the LC cell in electrically controlled birefringence (ECB) mode (see also 3.1) . Note the different LC molecules arrangements were achieved by using LC photoalignment layers. By selectively photopatterning the inner surfaces of the LC cells, one can make an LC cell act as a retarder, neutral density filter or rotator (Figure 4.48). The latter occurs wherever the orientation of the bottom inner layer is not identical to the top inner layer, in which case incoming light follows the rotation of the molecules. In E-field mode, the orientation of the LC molecules was parallel to the applied electric field (Figure 4.48).

The micropolarimeter array technology was fabricated comprising 40000 superpixels (200×200) [62,63]. The experiment has verified good LCMP performance in defining four Stokes vectors (Table 4.1) [62,63].

In contrast to previously reported micropolarizer arrays, the proposed implementation enables, for the first time, the extraction of circularly polarized components of light for a real-time full Stokes polarimetry. To accommodate applications with large oblique incidence angle, a 150nm ultra-thin visible-regime MWGP was used to reduce the overall thickness of the micropolarimeter array layer to 5μm. This provides a relatively good aspect ratio for pixel size larger than 10μm. Experimental results indicated that the LC-MWGP structure retained

MWGP's superior optical characteristics with a major principal transmittance of 75% and extinction ratios of around 1100.

Table 4.1. Comparison between experimentally extracted Stokes parameters and ideal values for different polarized inputs

Polarized inputs	Ideal values			Experimentally extracted			Error (%)		
	S_1/S_0	S_2/S_0	S_3/S_0	S_1/S_0	S_2/S_0	S_3/S_0	S_1/S_0	S_2/S_0	S_3/S_0
0° linearly	1	0	0	1.009	0.013	-0.009	0.9	1.3	0.9
90° linearly	-1	0	0	-1.012	0.017	0.012	1.2	1.7	1.2
45° linearly	0	1	0	-0.011	1.008	0.004	1.1	0.8	0.4
−45° linearly	0	-1	0	0.003	-0.998	-0.012	0.3	0.2	1.2
R-circularly	0	0	1	0.003	-0.013	0.977	0.3	1.3	2.3
L-circularly	0	0	-1	0.004	-0.003	-0.998	0.4	0.3	0.2

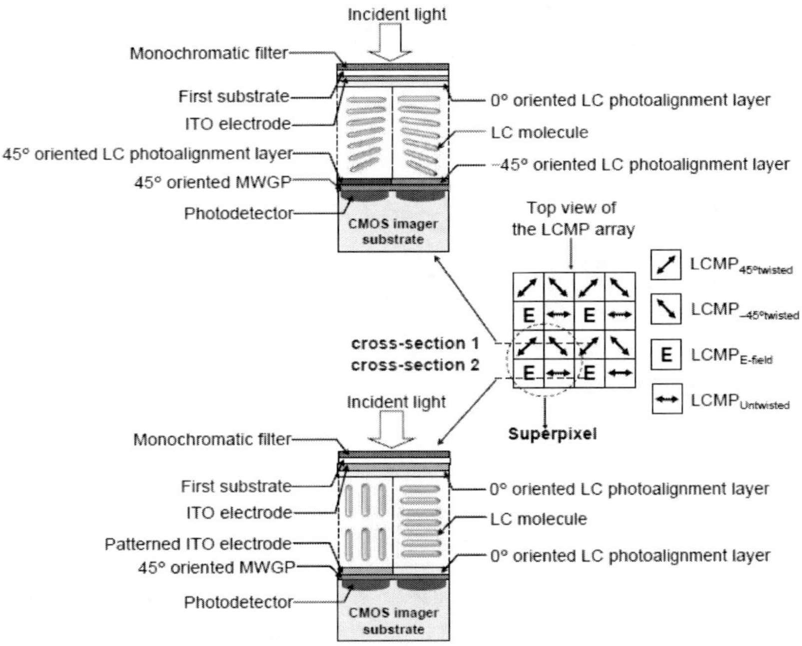

Figure 4.48. Top view and cross-sections of the proposed CMOS polarization image sensor's "superpixel" consisting of four LCMPs: LCMP 45°twisted, LCMP −45°twisted, LCMP E−field and LCMP Untwisted [62,63].

4.4.3. Other Types of LC Sensors

4.4.3.1. Liquid Crystal Sensors of Differential Pressure

The sensors of differential pressure based on different physical phenomena are of extremely wide usage in various branches of modern industry [14,59]. The main peculiarities of LC sensors which differ them from analogous devices are a high threshold sensitivity and electrically controlled characteristics [14]. In many cases a separation of a liquid crystal from a surrounding environment is needed. It can be done with the help of both limp and elastic diaphragms [14]. The possible approximate construction of the electrically controlled differential pressure sensor of the latter type is shown in Figure 4.49.

This technical decision realizes the mentioned above possibilities to control the technical parameters of LC sensors. Namely, the layer of a nematic liquid crystal of a variable thickness is used to get a wide dynamical range of the sensor. It also makes possible the visual registration of the applied pressure difference due to the sharp boundary separating the undistorted region of LC layer and the region where the orientational changes are strong enough to induce variation of LC local optical properties, i.e. birefringenece (3.15). The additional control of the boundary position (or the local optical response) is possible via application of electric voltage. In the construction shown in Figure 4.49 different voltages can be applied to the separate electrodes into the two parts of the cell. It provides a fine regulation of the optical response on the applied pressure difference.

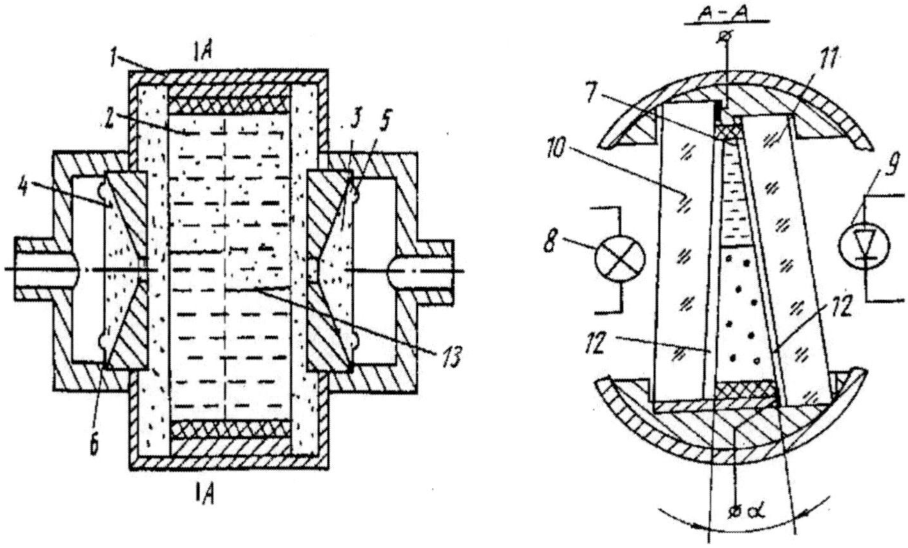

Figure 4.49. The possible construction of the liquid crystal sensor of pressure: 1- a body; 2- nematic liquid crystal; 3,4 - chambers; 5,6 - membranes; 7 - a capillary; 8 - a polarized light source; 9 -a photodetector; 10,11 - glass plates; 12 - current-conducting coating, 13 - a boundary [14].

The technical parameters of the differential pressure sensors are determined by a number of additional parameters. In particular, an elasticity of the membranes separating a liquid crystal and surrounding environment has to be small enough to transmit the pressure difference to the channel filled with LC. The best results were obtained for specially membranes made of a chemically stable rubber [14]. They provided the transmission of a pressure difference as small as 0.1 Pa at the membrane diameter of about 1 cm. The sensitivity of the sensor can be essentially increased via usage of membranes with larger diameters. Nevertheless, it also leads to increasing of the sensor sizes and LC amount. Other ways of optimization are essentially the same as those described above for the simpler LC sensors.

LC sensors provides elaboration of extremely sensitive and electrically controlled differential pressure sensors. It is possible to achieve the minimal value of the registered pressure difference of about 0.1 Pa which is two orders of magnitudes lower than for the most industrial sensors of comparable sizes used for low pressure measurements (see, for example, the sensor MPXV 5004, Motorolla, which provides measurements in the range 0...4kPa with 1.5% accuracy [14]).

4.4.3.2. Liquid Crystals for a Control of Liquid and Gas Flows

One of the perspective applications of the liquid sensors described above is a control of flows of liquid and gases [14]. Over 40% of all liquid, gas, and steam measurements made in industry are still accomplished using common types of differential pressure flowmeters [64]. In most cases it is needed to create an annular restriction in a pipeline with the help of orifice plate, Venturi tube, or nozzle to provide the pressure difference between the pressure upstream and pressure downstream of these elements. The increase in velocity at the restriction causes a corresponding increasing of the pressure difference. So, the measurements of the latter parameter can be used to determine the instantaneous rate of flow and the overall consumption in pipelines.

One of the major advantages of the differential pressure flowmeters is that the measurement uncertainty can be predicted without the need for calibration if it is manufactured and installed in accordance with one of the international standards covering these devices. In addition, this type of differential pressure flowmeter is simple, has no moving parts, and is therefore reliable. The main disadvantages of these devices are their limited range (typically 3:1), the permanent pressure drop they produce in the pipeline (which can result in higher pumping costs) and their sensitivity to installation effects (which can be minimized using straight lengths of pipe before and after the flowmeter) [14,64].

Liquid crystal differential pressure sensors seem to be very perspective for a control of flows due to a high sensitivity. In principle, it makes possible to avoid some disadvantages mentioned above. The general scheme of connection of the liquid crystal differential pressure sensor with a pipeline is shown in Figure 4.50 [14].

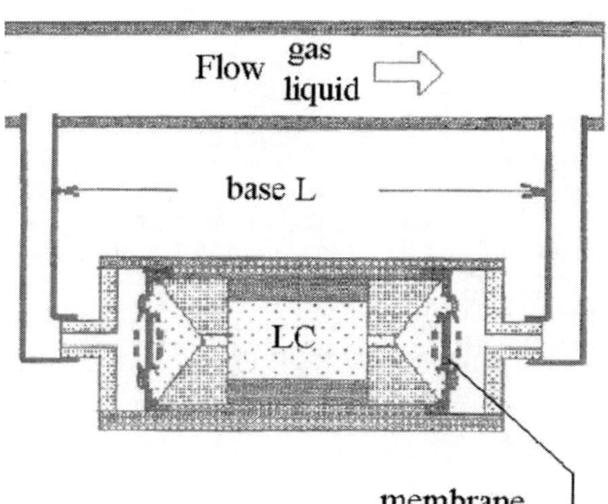

Figure 4.50. General scheme of application of LC differential sensor to control flows in a pipeline [14].

The pressure difference ΔP applied to the sensor can be produced in two ways. In case of laminar flows through the pipelines (of radius R) and of a constant cross section the viscous losses ΔP_{vis} play the dominant role. The can be expressed from Poiseuille formula

$$\Delta P = 8\eta_s LQ/\pi R^4, \tag{4.5}$$

where L and R are the length and the radius of the pipeline section, where the flow is controlled, Q is the instantaneous rate of the flow, η_s –shear viscosity of the flow [14]. It is interesting to estimate the viscous losses in flows of water as a liquid most important for practice. At R=1cm, L=10cm, $\eta_s = 10^{-3}$ Pa.s we will obtain the value of ΔP about 1 Pa for Q = 2 L/min. The corresponding Reynolds number Re equal to 2000 is close to the critical value $Re_c \approx 2300$ at which turbulence can arise in the flow of a liquid through the pipeline with the smooth walls. So, a very sensitive differential manometer has to be used to control the laminar flow of water in such pipeline. In principle, such device can be made on the base of the liquid crystal differential pressure sensors described above. As LC pressure sensors have a limited operating time a special decision has to be used at the control of one directional flow of a long duration. For example, it was proposed to provide periodical changes of the direction of the pressure gradient via usage of special unit as it is shown in Figure 4.51 [14].

The inversion of a direction of a pressure gradient is achieved with the help of a two-way valve. As it follows from equation (4.5), the control of laminar flows via registration if the viscous losses is suitable for pipelines of relatively small diameters (≤2cm in the case of water). In this case LC sensors show certain advantages in comparison with analogous devices [14,59,64] due to a high sensitivity to the pressure difference. In the case of turbulent flows through a pipeline of a large diameter the special inserts (orifice plates or conical Venturi tubes) have to be used (the example of the conical insert is shown in Figure 4.51). The description and technical characteristics of such units can be found in special literature. As usual, there is no simple connection between the pressure losses $\Delta P = \Delta P_{turb}$ and the instant flow rate similar to (4.5). In the first approximation: $\Delta P_{turb} \sim Q^2$ [14]. So, a wide dynamical range of the registered pressure has to be realized to provide a relatively narrow range of an instant flow rate. The efficient LC sensors can be achieved via usage of electric fields or (and) the cells of a variable gap.

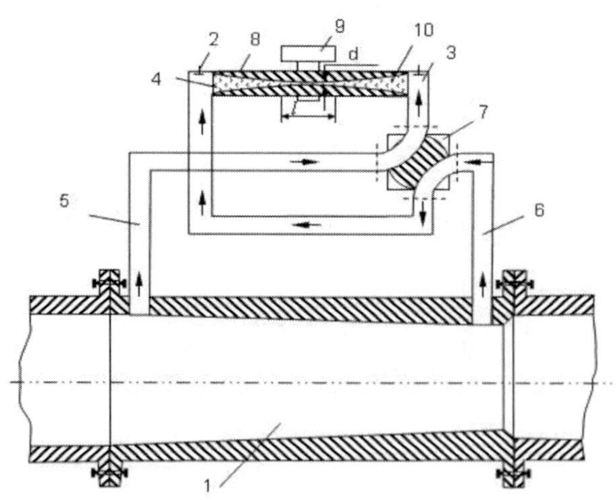

Figure 4.51. Scheme of usage of LC differential pressure sensor for a control of flows in pipelines. 1- Venturi tube, 2,3 – chambers with LC, 4 – membranes, 5, 6 – connected tubes; 7 - two-way valve, 8 – thin capillary with LC, 9 -photon-coupled pair, 10 – liquid crystal [14].

4.4.3.3. Various LC Sensors

A number of particular liquid crystal sensors (differential pressure sensors, sensors of acceleration, vibrations, inclination, liquid and gas flows) can be proposed on the base of the pressure gradient sensors [14]. The main advantages of such sensors in comparison with the analogous ones are high sensitivity and the possibility of electric control of the characteristics. Liquid crystal sensors based on the flow phenomena are most effective for registration of steady or low frequency mechanical perturbations. At the same time they can be used for registration, visualization and mapping of ultrasonic fields as the latter induce steady acoustical flows inside liquid crystal cells. In this case usage of electric (magnetic) fields is also effective for a control of the sensor characteristics.

A further development of a novel fiber optic low and high hydrostatic pressure sensing technique utilizing new classes of chiral nematic liquid crystals with a significantly reduced thermal sensitivity was proposed [65]. The low-pressure sensor (up to 1 MPa) was based on polarization effects caused by strong rotatory power of chiral nematics and pressure-induced deformations of a twisted nematic cell. The high-pressure sensor (up to 60 MPa) was based on intensity phenomena occurring in novel classes of chiral nematics with induced smectic phase. The method of hydrostatic pressure measurement offered a high response to pressure with a reduced temperature sensitivity. Potential areas of applications include pipelines and mining instrumentation, process-control technologies and environmental protection.

4.5. LC LIGHT CONVERTERS

4.5.1. Axially Symmetric Polarization Converter Using Patterned Liquid Crystal Quarter Wave Plate

The light with radial and azimuthal polarization could be very useful for advance optics, such as liquid crystal (LC) lens, polarization axis finder, beam shaping, optical trapping, laser cutting etc [66-68]. We describe the radial and azimuthal polarization converter, based on patterned liquid crystal quarter-wave plate, with high optical quality and relatively simple fabrication procedure. The properties of the radially and azimuthally polarized light can be described as in Figure 4.52. Considering polar coordinates (r, θ) with the polar axis L, the angle between polarization direction and the polar axis L is defined as φ at the point (r, θ), shown in Figure 4.52(a). Thus, the properties of the axially symmetric polarization can be described as follows:

$$\varphi(r,\theta) = q\theta + \varphi_0 \tag{4.6}$$

The constant q is the polarization order number, and φ_0 is the initial polarization orientation for $\theta = 0$. When $q = 1, \varphi_0 = 0$, the light is radially polarized, which is shown in the Figure 4.52 (b). When $q = 1, \varphi_0 = \pi/2$, the light is azimuthally polarized (Figure 4.52 (c)).

It is important to discuss the optical properties for circularly polarized light passing through the unpatterned quarter-wave plate. If the optical axis of the quarter wave plate is along the x axis, as shown in Figure 4.53, and the input light is left (Figure 4.53(a)) or right (Figure 4.53(b)) handed circularly polarized light.

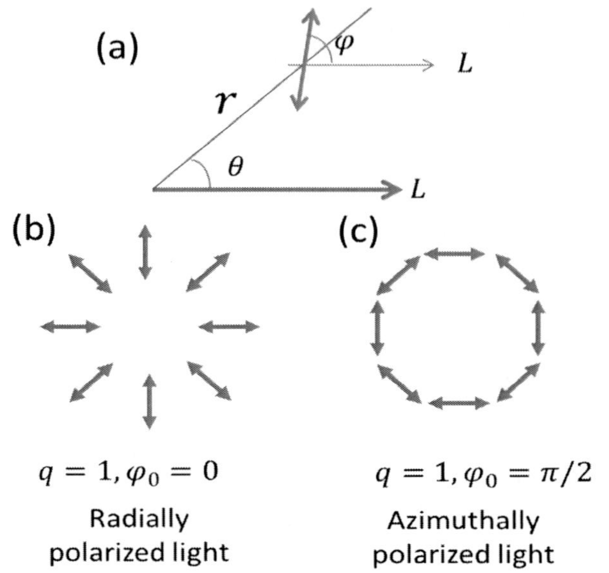

Figure 4.52. Description of radially and azimuthally polarized light. (a) polar coordinates model (b) radially polarized light, (c)azimuthally polarized light [66].

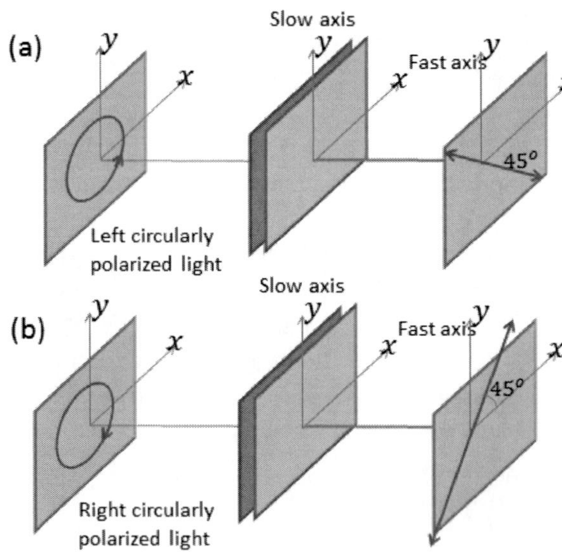

Figure 4.53. (a) Left handed and (b) right handed circularly polarized light passing through the quarter wave plate [66].

Then the output light property can be described by Jones vector in Eq. (4.7) (for input light with left circular polarization) and Eq. (4.8) (for input light with right circular polarization):

$$\overrightarrow{E_{L\_out}} = \begin{pmatrix} 1 & 0 \\ 0 & \exp(-i\frac{\pi}{2}) \end{pmatrix} \frac{1}{\sqrt{2}} \begin{bmatrix} 1 \\ -j \end{bmatrix} = \frac{1}{\sqrt{2}} \begin{bmatrix} 1 \\ -1 \end{bmatrix} = \begin{bmatrix} \cos(-\pi/4) \\ \sin(-\pi/4) \end{bmatrix} \quad (4.7)$$

$$\overrightarrow{E_{R\_out}} = \begin{pmatrix} 1 & 0 \\ 0 & \exp(-i\frac{\pi}{2}) \end{pmatrix} \frac{1}{\sqrt{2}} \begin{bmatrix} 1 \\ j \end{bmatrix} = \frac{1}{\sqrt{2}} \begin{bmatrix} 1 \\ 1 \end{bmatrix} = \begin{bmatrix} \cos(\pi/4) \\ \sin(\pi/4) \end{bmatrix} \quad (4.8)$$

Thus it is clear that the output light is linearly polarized light with angular shift of $\pi/4$ from the fast optical axis of the quarter-wave plate in clockwise direction (for input light with left circular polarization) and in anticlockwise direction (for input light with right circular polarization).

Therefore, when circularly polarized light passes through a quarter-wave plate with fast optical axis spatially variable, the output light will be linearly polarized light with its polarization direction spatially shifted from the fast axis with the angle of $\pi/4$. Thus, for generating the radially or azimuthally polarized light, one can simply design the patterned quarter-wave plate with its fast axis $\pi/4$ shifted from the target polarization direction. The fast optical axis of the quarter-waveplate can be denoted as:

$$\alpha(r,\theta) = q\theta + \alpha_0 = q\theta + \varphi_0 \pm \pi/4 \quad (4.9)$$

so, to construct the axially symmetric polarization light with $q=1, \varphi_0 = 0 \; or \; \pi/2$, the axially symmetric quarter wave plate with $q=1, \alpha_0 = \pi/4$ can be used. When the input light is left or right circularly polarized, the output light will be radially or azimuthally polarized, as shown in Figure 4.54 (a) and (b) respectively.

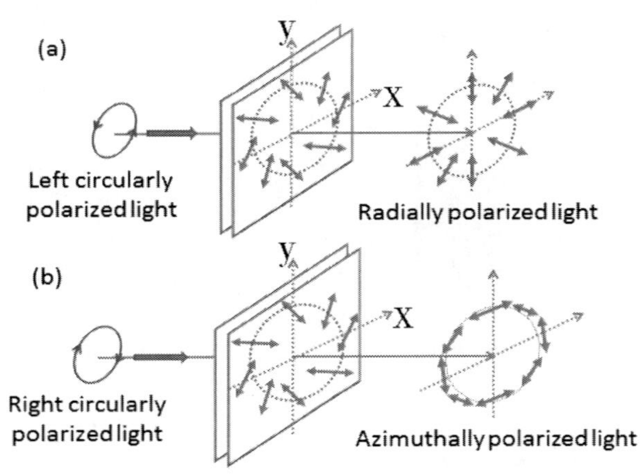

Figure 4.54. Patterned quarter wave plate with property of $q=1, \alpha_0 = \pi/4$ working as (a) radial polarization converter and (b) azimuthal polarization converter [66].

Therefore, a patterned quarter-wave plate with axial symmetry of $q=1, \alpha_0 = \pi/4$ can be used as both, radial and azimuthal polarization converter. The output light with radial polarization or azimuthal polarization can be determined by input light with left or right handed circular polarization.

To fabricate the LC patterned quarter wave plate with axial symmetry with $q=1, \alpha_0=\pi/4$ as shown in Figure 4.54, the most efficient way is to control the alignment layer [66-68]. Usually, the fast axis of LC wave plate is determined by the azimuthal angle of the alignment. The traditional method of preparing alignment layer is using rubbing, and the azimuthal angle will be the same as the rubbing direction. However, for the alignment with azimuthal axial symmetry $q=1, \alpha_0=\pi/4$ rubbing method is not feasible because of its fabrication complexity. The photo-alignment is a fast developing technology which can decrease alignment defects and is relatively simple for multi-domain alignment in the comparison with rubbing method. For the photo-alignment, linearly polarized light in UV or visible range is used to expose the alignment layer. The azimuthal angle can be the same or perpendicular to the exposing light polarization, depending on the chemical properties of the photoalignment material. The glass substrate was fixed on to a rotating stage as in the experimental setup shown in Figure 4.55. A UV lamp (Omnicure S1000) was used to illuminate the photoalignment layer of SD1 sulfonic dye [66-68]. Two lenses were introduced in the light path to broaden the beam size. A line mask with a cylindrical lens was used to project a narrow light beam onto the substrate. Then after a wire-grid polarizer with axis at $\pi/4$ to the x-axis (Figure 4.55) was placed in the light path to provide linearly polarized light. The stage with SD1 substrate was rotated at the speed of one degree per minute. Thus, the half circle of rotation (i.e. 180º) of the SD1 substrate provides the alignment, which gives the azimuthal angle with the axial symmetry of $q=1, \alpha_0=\pi/4$. Then after, the liquid crystal polymer (LCP) UCL 017 (DIC, Japan) was coated on the top of the SD1 layer, thus the constructed layer thickness of LCP gives the retardation equal to the quarter-wave for wavelength 450nm. The SD1 layer will align the LCP spatially with the same axial symmetry. Then the substrate was exposed to UV light to polymerize and fix the orientation of the LCP and thus, the polarization converter was fabricated [66-68].

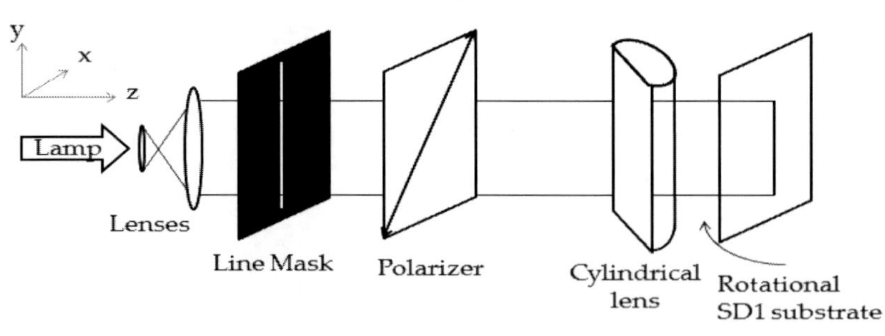

Figure 4.55. Experiment setup for the patterned photo-alignment [66-68].

The results are shown in Figure 4.56. In Figure 4.56 (a), the polarization converter was placed between crossed polarizers. We can see that when the fast optical axis is the same with either direction of polarizers, it shows black state. On the other hand, if the incident light is left and right handed circularly polarized light, then after passing through the polarization converter, the output light becomes radially polarized light (Figure 4.56 (b)), and azimuthally polarized light, (Figure 4.56 (c)). An analyzer was placed after the polarization converter to monitor the output light property. It could be seen that when light polarization was at the same direction as the analyzer, the transmittance was zero, which showed black in the figure. Therefore, it confirms the working of proposed polarization converter.

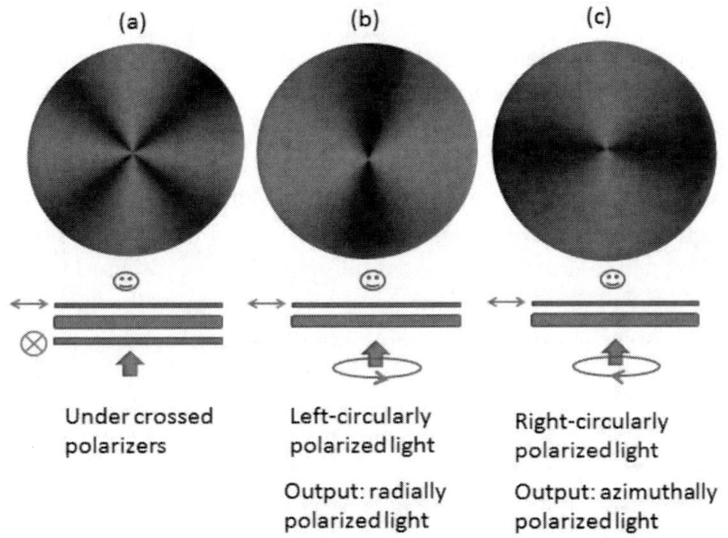

Figure 4.56. Optical photographs and illustrations of the experimental results for the polarization converter under different condition [66].

Thus, the proposed polarization converter, depending on the incident light polarization, can be used to manipulate the polarization plane of the output light in several ways [66]. If the incident light is linearly polarized, a uniform quarter-wave plate which has fast optical axis at $\pi/4$ with the linear polarization direction can generate circular polarized light. On the other hand, if the incident light is unpolarized, a linear polarizer with a uniform quarter-wave plate can be used to generate circularly polarized light. In addition to this, the patterned quarter-wave plate with axial symmetry of $q=1, \alpha_0=\pi/4$ has an important property, i.e. if the substrate is inversed, the quarter-wave plate will have the axial symmetry of $q=1, \alpha_0=-\pi/4$. In this case, for left circularly polarized light, the patterned quarter-wave plate will act as an azimuthal polarization converter and for right circularly polarized light it will act as a radial polarization converter.

In comparison with the axial polarization converter using twisted nematic, the patterned quarter wave plate axial polarization converter has no defect line which appears in twisted nematic cell at the boundary of opposite twist. This type of polarization converters needs only half of the retardation than the half wave plate polarization converters, consequently only half of the film thickness is needed, which results in less dispersion.

Thus a radial and azimuthal polarization converter is shown using patterned liquid crystal quarter-wave plate [66]. Axially symmetric photoalignment was used to fabricate the patterned quarter-wave plate. The radially and azimuthally polarized output light can be generated with input light with random, circular or linear polarization. Therefore, the proposed device has immense potential for application in optical trapping systems, to increase the trapping force, in laser cutting machine, as well as to increase the cutting speed and several other electro-optical and telecommunication components.

4.5.2. Tunable Liquid Crystal q-Plates

The device, called "q-plate" (QP) is essentially a birefringent wave plate with inhomogeneous patterned distribution of the local optical axis in the transverse plane [67]. The pattern of the optical axis distribution is defined by semi-integer "topological charge" q. When an input light beam traverses the q-plate, a $2q$ amount of orbital angular momentum (OAM) is transferred into the beam, with the sign determined by the input polarization helicity. The efficiency of this conversion is given by the optical retardation δ of the QP, that can be controlled by temperature or electrical field, in case of LC-based QPs. Such device has already found a number of applications in various fields of optics [66-68]. Polymer based QPs were manufactured, however such QPs, being solid and non-tunable, are suitable for single wavelength, require high level control of the sample thickness during the manufacture and cannot be used in the applications, where the conversion efficiency must be changed or switched. The photoalignment technology allows, in principle, to create any topological semi-integer charge and does not need precise control of the cell gap of the sample.

The orientation of the local optical axis, in case of the axial dependence only can be given by (4.6) $\alpha(\varphi) = q\varphi + \alpha_0$, where q is the topological charge, φ is the transverse azimuthal coordinate and α_0 is a real number. In case of $q = 1$ the LC forms a circularly symmetric azimuthal or radial distribution of the optical axis. While the $q = 1$ QPs can be manufactured by rubbing, other patterns are not realizable with this method.

The photoalignment technique was applied to manufacture LC QPs using the set-up shown in Figure 4.55 [66]. The SD1 azo-dye provides planar alignment for the LC in the direction perpendicular to the incident polarization light, with anchoring energy comparable with the polyimide rubbing based alignment. By adjusting the ratio between angular speeds and relative direction of the rotation of the two motors, different topological charges were possible to achieve. For example, rotating both sample polarizer with the same speed but opposite direction will induce charge $q = 2$ into the cell. After the exposure, the samples were filled with the LC (MLC 6080 mixture from Merck). An exposure time of 2 hours was enough to provide high quality alignment of the LC layer. A subsequent heating of the sample to the temperature above the LC clearing point and slow cooling helped to remove occasional LC alignment defects. Topological charges $q = 0.5$, 1.5, 3, as shown in Figure 4.57, were realized, however any semi-integer charge can be realized with such technique [67].

The LC q-plates efficiency can be controlled by changing the phase retardation, changing the temperature of the cell or by applying electrical field [67]. QPs were made with glass substrates that had ITO conducting coating, allowed us to apply electrical field to the LC layer. The normalized with respect to the total output power measurements dependence from the voltage applied was also measured. The circularly polarized input beam was sent into the QP performing a polarization conversion and after that the converted and non-converted part were separated spatially by a quarter-wave plate and a polarizing beam splitter. The measurements were done for 543.5 nm and 633 nm input light wavelengths. The normalized with respect to the total output power measurements are shown in the Figure 4.58. The maximum conversion efficiency was found to be more than 99% [67].

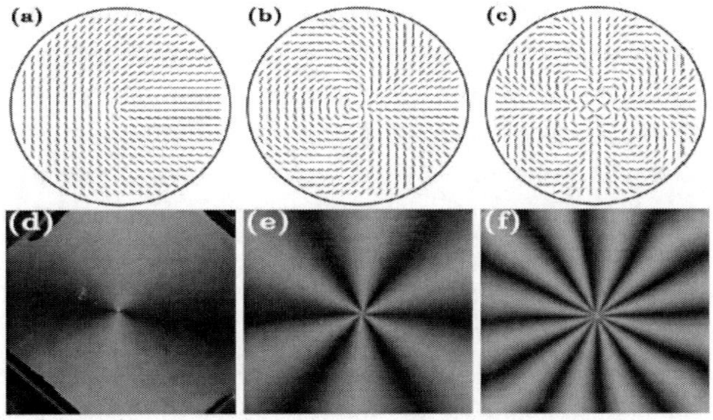

Figure 4.57. Examples of the LC patterns that correspond to different topological charges and the photos of the samples prepared under crossed polarizers. The simulated charges are (a) – q = 0:5, (b) – q = 1:5, (c) – q = 3 and (d), (e) and (f) are the corresponding q-plates [67].

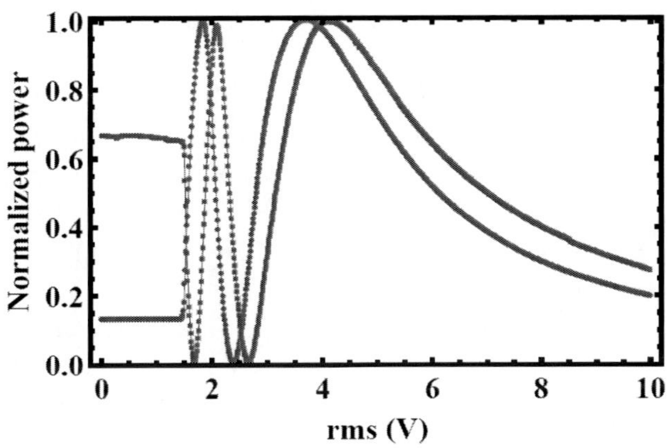

Figure 4.58. Converted intensity of the input beam as a function of the voltage, applied to the LC cell. Red line - 633 nm input beam wavelength, green line - 534.5 nm input beam wavelength. The converted power values were normalized with respect to the total light power output from the q-plate [67].

Thus a specific electrically tunable q-plated were made using azo-dye SD1 photoalignment materials. The application to optical data processing (polarization conversion) was successfully demonstrated with the efficiency of 99% [67].

4.5.3. Liquid-Crystal Spatial-Mode Converters for the Orbital Angular Momentum of Light

During the last years, the orbital angular momentum of light (OAM) has attracted a growing attention in the scientific world and is finding more and more applications in various fields of optics. While the polarization state of the photon can be easily controlled by polarizers, wave-plates and electro-optical phase retarders, for the OAM space there are not yet such convenient devices. Various techniques of OAM state generation have been introduced up to now, such as holograms (including reconfigurable ones, made using spatial light modulators), spiral phase plates, q-plates, various interferometric setups, and others. A liquid-crystal Pancharatnam-Berry optical element (PBOE) was demonstrated [68]. A Pancharatnam-Berry optical element is a space variant phase element that exploits the geometrical phase arising in polarization manipulations for reshaping the optical wavefront. PBOE was realized as a birefringent waveplate of uniform phase retardation δ, whose angle α between slow optical axis and the x axis of the fixed laboratory reference frame is not constant, but is described by a prescribed function $\alpha = f(\rho, \varphi)$, where where ρ and φ are the polar coordinates in the waveplate transverse plane. In other words, a fraction of the incident circularly polarized light undergoes a helicity inversion and gains an additional phase factor of $\pm 2f(\rho, \varphi)$, where the sign depends on the input polarization helicity. This additional phase factor is not due to refractive index change or non-uniform thickness of the optical element, but due to the geometrical phase arising from the space-variant polarization manipulation. The efficiency of the beam conversion depends on the phase retardation δ and is maximum for the half-wave $\delta = \pi$ condition. Some examples of these optical axis patterns are given in Figure 4.59 [68], together with images of corresponding fabricated samples. The ease of use and compactness of our device makes it still interesting and suitable for certain OAM manipulation applications.

A nematic LC was used to fabricate our PBOE devices [68]. The desired planar alignment of the LC was induced using a photoalignment technique. The scheme of our fabrication setup is shown in Figure 4.55 and described above. A total exposure of 1.5 h was enough to provide high quality LC alignment. PBOE devices were realized in this work [68]. As in the case of q-plates, or other LC cells, the birefringence, and thus the phase retardation δ in a phase plate was controlled by applying an external AC electric field (see also Figure 4.58). Typical maximum and minimum efficiencies, switching times and overall tuning behavior of our devices were found to be identical to those of q-plates. As a main test of our PBOEs, the wavefront transformation of a laser beam passing through the sample was studied using a Mach-Zehnder interferometer (MZI) [68]. In one of the arms of the MZI the input beam was manipulated by means of q-plates. The other arm was used as reference beam. By blocking the reference arm it was possible to record the intensity pattern of the analyzed beam, too. The interference and intensity images of input and output beams are shown in Figure 4.60 below.

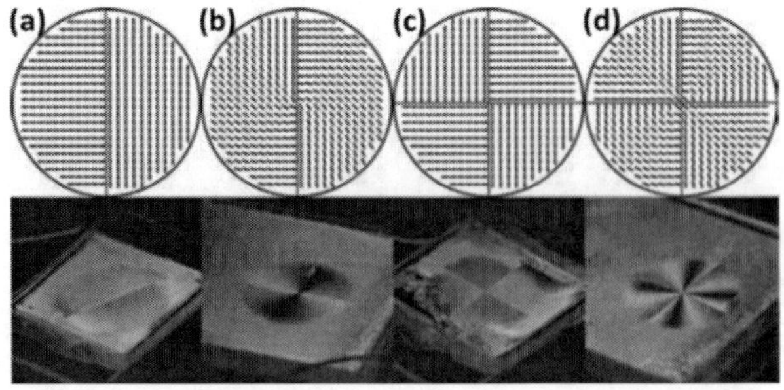

Figure 4.59. Top row: optical axis orientation patterns of various shapes. Bottom row: corresponding fabricated samples seen between crossed polarizers under oblique illumination. Different colors correspond to different optical axis orientations, with dark areas corresponding to the zones where the optical axis is oriented parallel to one of the polarizers [68].

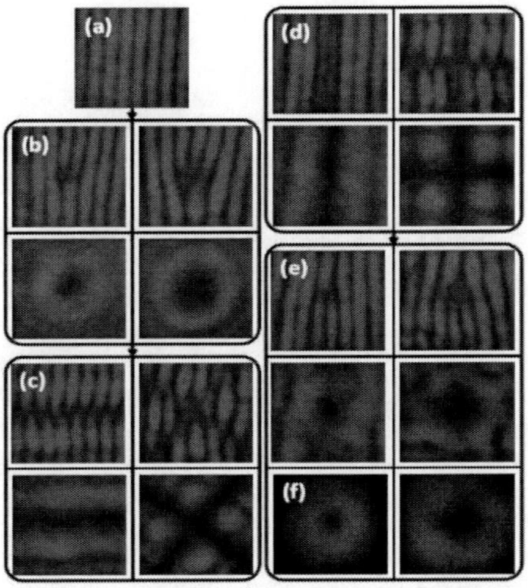

Figure 4.60. Interference and intensity images of input and output beams. Left column: (a) - input beam. (b-f) -variuos output beams modified by PBOEs [68].

Thus LC optical element that can operate by introducing a polarization-dependent phase-only transformation of the input beam was demonstrated [68]. It is worth noting that the devices introduced here can be also used in the single photon quantum regime. The device may find application in all fields in which a complete manipulation of the light orbital angular momentum (OAM) is needed, such as, for instance, optical communication, optical computing, and quantum cryptography.

4.5.4. Arbitrary pattern Fabrication in Liquid Crystal (LC) Alignments and Local Polarization Control for Light Wavefront

To obtain arbitrary LC microstructures, a digital mirror device (DMD) based micro-lithography system was developed as shown in Figure 4.61 [69].

It consists of several sub-systems, including a light emission part, a dynamic pattern generation part, an image focusing part and a monitor part, all of which function in cooperation to provide correct exposure on the substrate. A mercury lamp filtered at 320-500 nm along with a collimating lens provide a uniform and collimated light beam. The beam illuminate onto the DMD surface through an iris. DMD (1024 × 768), as a dynamic mask, consists of more than 786,000 micro-mirrors. Independently tilted by an electrostatic force, each mirror (13.68 μm × 13.68 μm in size) can be switched between “on” and “off” states by toggling the applied voltage. The bundle of light reflected by the “on” state mirrors will form a desired light pattern. Then the pattern was focused by an apo-chromatically corrected projection lens and recorded on the substrate placed at the image plane. The light reflected by the substrate is then collected into a CCD, which is used to monitor the focusing of the image. In this system, several parameters are adjustable: the focal length of the collimating lens, the distance between the collimating lens and DMD and that between DMD and the objective lens. These parameters were optimized thus achieved a uniform illumination on DMD and a clear image at the substrate [69].

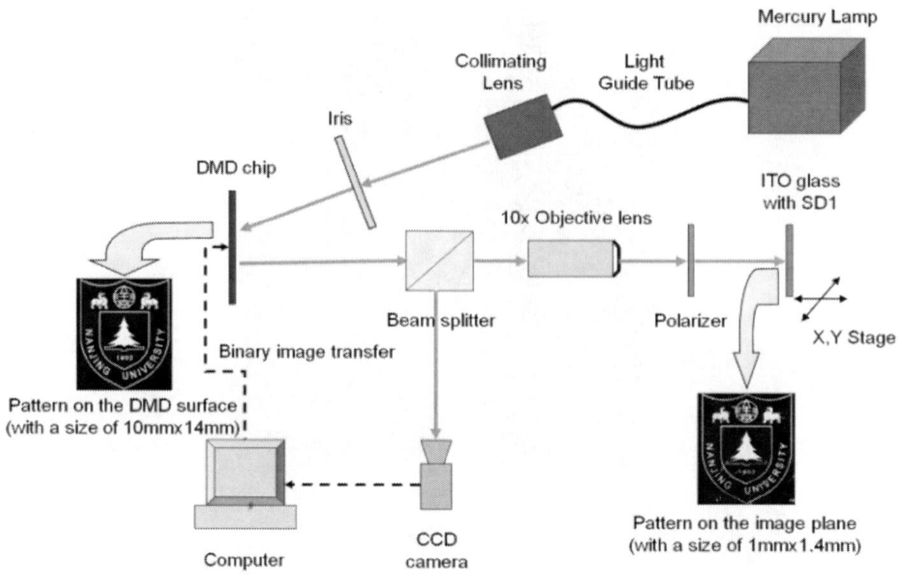

Figure 4.61. Schematic illustration of DMD based micro-lithography system [69].

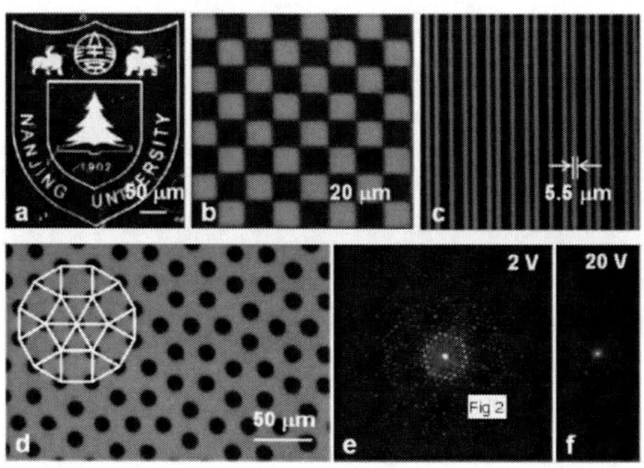

Figure 4.62. Micrographs of different LC patterns under a cross-polarized microscope: (a) logo of Nanjing University, (b) chequerboard, (c) Fibonacci grating, (d) 12-fold symmetric quasi-periodic structure and (e-f) diffraction patterns of the sample shown in (d) under different applied voltages. All cell gaps are 6 μm [69].

Two ITO-coated glass substrates are UV-Ozone cleaned and spin-coated with SD1 films. The two substrates were both uniformly irradiated by linearly polarized light to create a photoalignment in the same direction with a dose of ca. 5 J/cm^2. One of the substrates was then placed at the image plane of the system and exposed again with a desired pattern protracted by CAD or Photoshop software. Herein the light polarization was set different from that of previous exposure. SD1 [35] shows excellent alignment rewrite-ability (see also 3.6.3), therefore the exposed regions were realigned while the mask shadow domains still keep the previous alignment. An LC cell is assembled by the two substrates and then filled with LC mixture E7 at isotropic phase. After cooling to room temperature, the LC molecules show patterned orientations due to different alignment directions among adjacent domains. Several kinds of phase patterns using this system are presented in Figure 4.62.

The bright regions show twisted nematic (TN) alignment of LC while the dark regions show planar homogeneous (PA) alignment. All samples are observed by a cross-polarized microscope with polarizer direction parallel to PA direction. Figure 4.62(a) shows an LC cell patterned with the logo of Nanjing University [69]. As one can see, the brightness of different parts is uniform and the edge of the pattern is sharp. The line width of the frame measures 11 μm, well coordinate with our design (8 pixels on DMD). Figure 4. 62(b) presents a 2D chequerboard grid. In spite of slight distortion due to the aberration of projection lens, the pattern keeps square. No beam expansion exists and remarkable improvement is achieved in pattern quality. Moreover, 1D and 2D quasi-periodic LC patterns have also been fabricated. Figure 4.62(c) exhibits a Fibonacci structure with feature size of 5.5 μm (4 pixels on DMD). In Figure 4.62(d), a 12-fold symmetric pattern is presented. From the diffraction patterns under different voltages shown in Figure 4.62(e-f), one can easily figure out 12-fold rotation axes that would never occur in periodic structures. This is the first time that LC quasi-periodic patterns were realized. All above results prove a faithfully replication ability of the micro-lithography system. Therefore arbitrary structure fabrication could be achieved with this system as the free pattern design on DMD.

Besides flexible pattern manipulation, local alignment direction control was also realized by rotating the polarizer. A polarization rotator array with different angles in a single LC cell was demonstrated. The two substrates were photoaligned in parallel and then 9 regions on one of the substrates are realigned sequentially through changing the DMD pattern (disk position) as well as the polarization of activating light. Thus, 9 TN disks with twist angles varying from 10° to 90° at an interval of 10° were formed, observed as different gray scales in Figure 4.63(a) and (b).

During observation, the light is incident to the unpatterned substrate first, with its polarization parallel to the alignment direction, and then when passing through the cell, only components along the analyzer directions could be observed. The results suggest the possibility of arbitrary polarization control of any part of the wavefront, which might be used in vector beam generation. Furthermore, in the whole exposure process, no mechanical movements of either DMD or substrate are needed, avoiding the registration problem of conventional multi-step lithography.

Thus a technique for arbitrary pattern fabrication in liquid crystal (LC) alignments and local polarization control for light wavefront was proposed and implemented [69]. A micro-lithography system with a digital micro-mirror device as dynamic mask forms arbitrary micro-images on photoalignment layers and further guides the LC molecule orientations. Besides normal phase gratings, more complex 2D patterns such as quasicrystal and checkerboard structures are demonstrated.

Compared to other techniques, the method enables the arbitrary and instant manipulation of LC alignments and light polarization states, facilitating wide applications in display and photonic fields [69].

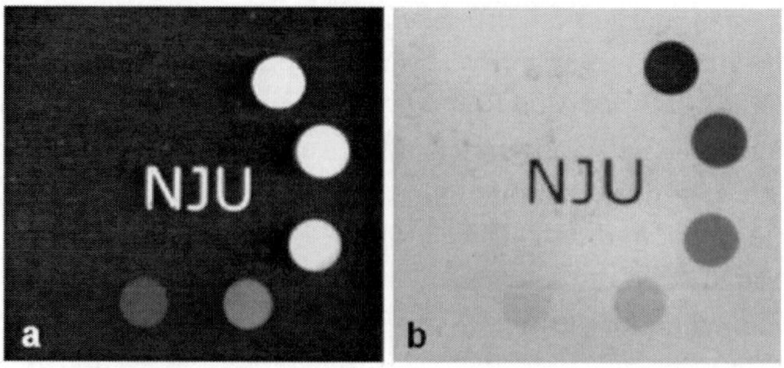

Figure 4.63. Micrographs of LC polarization rotators with different twist angles under a polarizing microscope with (a) crossed and (b) parallel polarizers [69].

4.6. Field Sequential Color (FSC) Displays

Field sequential color (FSC) displays is becoming popular as a method of eliminating the color filters in LCDs (Figure 3.76). In LCD panels, the color filter accounts for some 25% of the total cost of the LCD display panel. Thus, eliminating the color filters represents a significant way of reducing costs and simplifying the production process. FSC displays will considerably reduce the LCD price and promote further improvements for brightness, resolution and color gamut. The LCD controller in a color sequential system is synchronized with the backlight so that when a given color backlight is on, only the matching color sub-pixels in the LCD are turned on. The end result is color that is much richer and more realistic. Eliminating the color filters in LCD panels has a further advantage. Red, green, and blue color filters used on sub-pixels absorb up to some 70% of the light output from the backlight - thus eliminating the color filters. However, eliminating the color filters and getting the color subsequently in time one should be able to solve another problem: the LCD switching time should be considerably fast. The driving of such color sequential LCD includes three sub-frames. The response time of the LCD needs to be very fast in order to fit the field sequential driving method. Suppose that the frame frequency is 60 Hz, i.e. the frame duration is 16.67 ms. Each sub-frame will have 5.5ms. Again, each sub-frame has to further divide into three parts. The first one is the data loading time which is about 0.5ms. Secondly, it comes to the LC response time which is about 2.1 ms. Finally, the total LED illumination time is around 2.9 ms. However, typical response rate of LCD is about 240 frames per second and increases twice, when 3D time sequential addressing is used, so the LC response time should be about 8 times shorter, i.e. about 250 μsec, which is really a challenge.

4.6.1. LC Advantages among Organic Light Emitted Diodes (OLED), Plasma Display Panels (PDP) and Micro-Electro-Mechanical (MEM) devices. Fast Nematic and Cholesteric FSC LCD

The competing technologies of liquid crystal FSC display devices are mainly based on the applications of organic light emitted diodes (OLEDs) without color filters and digital light projectors (DLPs) [70].

OLEDs can provide a high brightness, wide color gamut and wide viewing angles. However, OLEDs without color filters suffer a comparatively low resolution and certain technological problems. Each OLED pixel need to be driven by two TFT transistors, which considerable decreases the yield and increases the price (Figure 4.64).

Digital light projectors (DLPs) are based on micro-electro-mechanical (MEM) mirrors, which are characterized by only two states (bright and dark), so a subsequent time sub-frames are required for fast moving images to get the gray scale levels required for extended color reproduction. The latter limitation considerably decreases the color triangle and exhibit a color break–up (blur) which increases the power. Besides the MEM resolution is limited to 10-15 microns for a practical implementation as the mechanically rotating mirror cannot be of a smaller size (Figure 4.65).

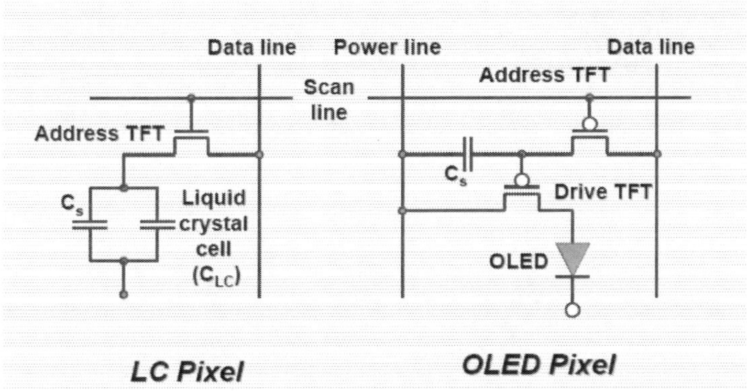

Figure 4.64. Basic driving structure of LC pixel (left) and OLED pixel (right) [70].

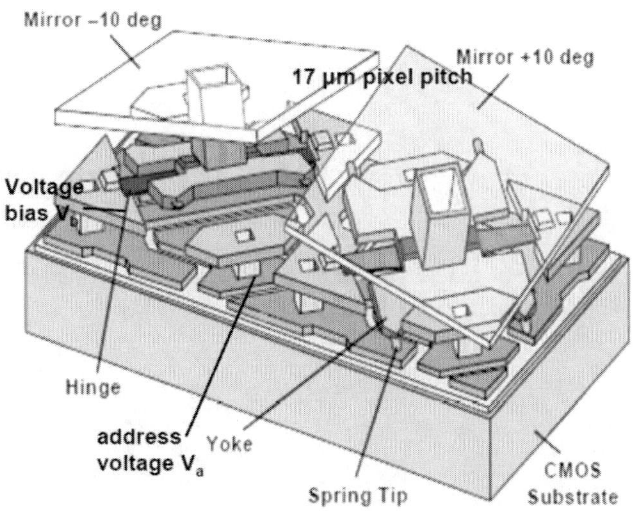

Figure 4.65. Digital light projectors (DLP) based on micro-electro-mechanical (MEM) mirrors [70].

Fast electrooptical modes in nematic LC, such as optically compensated birefringence [71] (OCB, Figure 4.66)), no bias bend [72] (NBB, Figure 4.67) or stressed splay twist [73] (SST, Figures 3.21) and total internal reflection [74] (TIR, Figure 4.3) modes do not seem to be very helpful to meet the new requirements for FSC and 3D display applications as the switching time is still in a millisecond range [70].

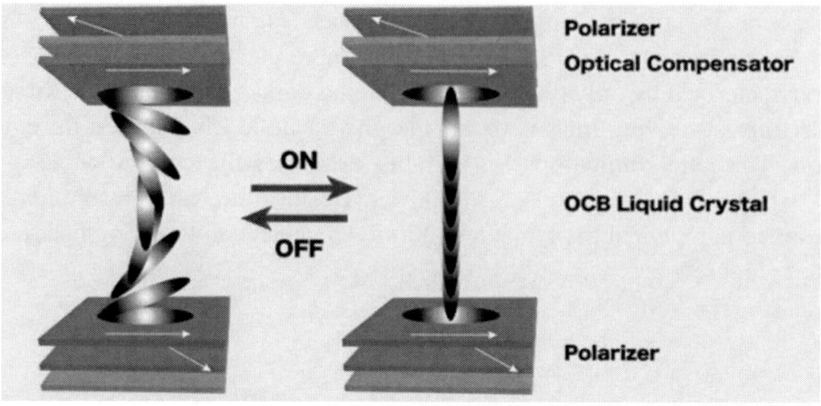

Figure 4.66. Optically compensated (OCB) mode, see also (3.18) [71] .

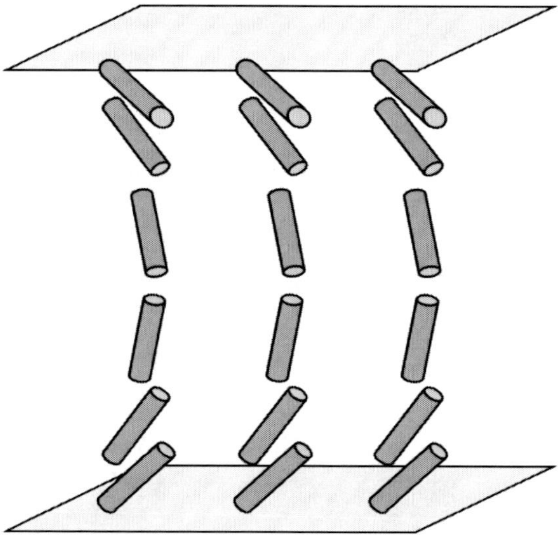

Figure 4.67. No bias bend (NBB) mode. The initial splay state is not created by electric field like in OCB mode (Figure 4.66), but using the high tilt angle at surface [72].

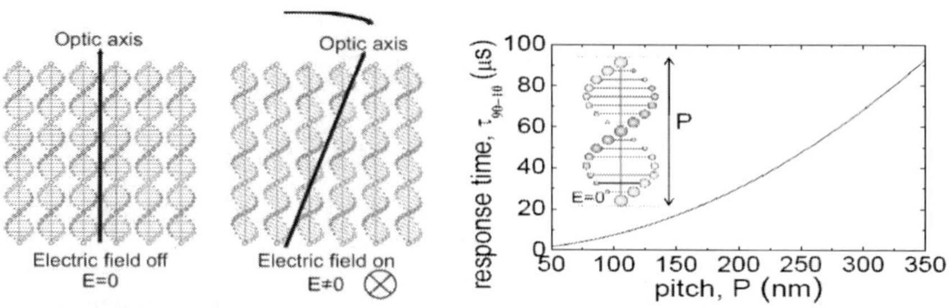

Figure 4.68. Flexoelectric effect in chiral LC mixtures. The deformation of the helix (see also Figure 3.30), which is equal to the rotation of the helix axis (left) and the characteristic switching times [75].

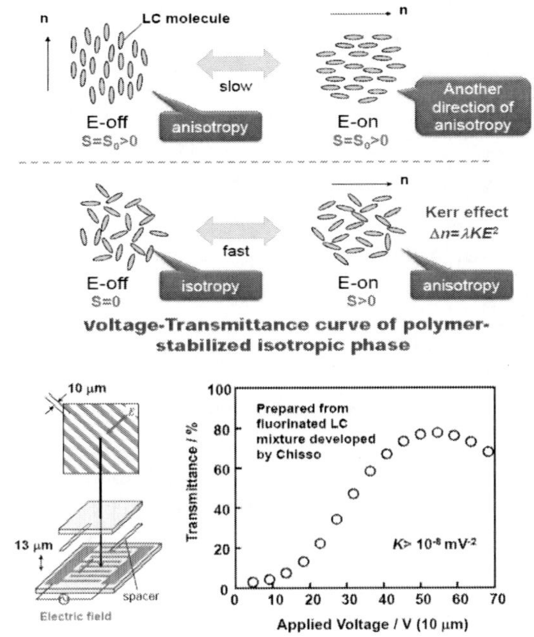

Figure 4.69. LC switching in blue phase [76].

Fast electrooptical modes in cholesteric LC, such as flexoelectric effect (Figure 4.68, Figure 3.30) and blue phase (Figure 4.69) modes seem to be very helpful to meet the new requirements for FSC and 3D display applications, but further work is needed because of several fundamental problems such as too high applied electric field, hysteresis and alignment problems, non-sufficient transmission level.

4.6.2. Ferroelectric FSC LCD

FLCD is the fastest LC mode, which can work with both fast responses time and low driving voltages and is highly suitable for field sequential (FSC) applications. Fast switching ferroelectric liquid crystal (FLC) displays (FLCD) is a good candidate for the new generation of field sequential color (FSC) LCD, which proved to be better in response time, than usually used nematic LC. The best FLC parameters can be obtained on the basis of ESH mode (3.5.2.5, Figures 3.44-3.46).

Novel photoaligned FLC devices may include FSC FLC with a high resolution, low power consumption and extended color gamut, which can be used in the screens of portable PCs, mobile phones, PDAs. The switchable goggles and lenses based on new FLC prototypes can be efficiently applied in the new generations of switchable 2D/3D LCD TV. FSC FLC microdisplays, which is now one of the most advanced technologies for pico-projectors can be also made on the basis of new materials and electrooptical modes in FLC shown above in 3.5. The photoalignment technology enables to solve the key problems usually faced in FLCD applications, such as (i) quality of FLC alignment on sufficiently large surface area; (ii) appropriate adjustable anchoring energy and pretilt angle; (iii) low losses in the alignment layers due to their small thickness etc. [35].

The further the development of novel photoaligned fast FSC FLCD is aimed at: (i) further fundamental study of the new appropriate electrooptical modes used for switching; (ii) further understanding of physical mechanisms of FLC interaction with a photoaligned surface of different photosensitive nature to produce a stable alignment with a controllable anchoring energy and pretilt angle over sufficiently large surface area; (iii) development of new fast responded FLC materials with fast switching and a sufficient number of switchable gray levels (V-shape switching); (iv) implementation of the working prototypes of novel FSC FLC displays; (v) investigations of regimes of operation to allow to use efficient addressing of FLCD.

REFERENCES

[1] Vladimir G. Chigrinov, Liquid crystal applications in photonics, *Frontiers of Optoelectronics in China*, Vol.3, 1674-4128 (2010).

[2] http://computer.howstuffworks.com/fiber-to-the-home.htm

[3] http://www.lightwaves2020.com

[4] http://www.vescentphotonics.com

[5] Beresnev, L; Haase, W. "Ferroelectric Liquid Crystals: development of materials and fast electrooptical elements for non-display applications", *Optical Materials*, 9, 201 (1998).

[6] Wang, Qian; Farrell, Gerald. Integrated liquid-crystal switch for both TE and TM modes: proposal and design, *J. Opt. Soc. Am. A*, 24, 3303 (2007).

[7] Crossland, W; Wilkinson, T; Manolis, I; et al. "Telecommunications applications of LCOS devices", *Mol. Cryst. Liq. Cryst.*, 375, 1 (2002).

[8] Crossland, WA; Clapp, TV; Wilkinson, TD; Manolis, IG; Georgiou, AG; Robertson, B. Liquid Crystals in Telecommunications systems, *Mol. Cryst. Liq. Cryst.*, 413, 363 (2004).

[9] Beccherelli, R; Bellini, B; Donisi, D; d'Alessandro, A. Integrated optics devices based on liquid crystals, *Mol. Cryst. Liq. Cryst.*, 465, 249 (2007).

[10] Zhang, A; Chan, KT; Demokan, MS; Chan, V; Chan, P; Kwok, HS; Chan, A. Integrated liquid crystal optical switch based on total internal reflection, *Appl. Phys. Lett.*, 86, 211108 (2005).

[11] Hoshi, R; Nakatsuhara, K; Nakagami, T. Optical switching characteristics in Si-waveguide asymmetric Mach-Zehnder interferometer having ferroelectric liquid crystal cladding, *Electr. Lett.*, 42, 635 (2006).

[12] Riboli, F; Daldosso, N; Pucker, G; Lui, A; Pavesi, L. Design of Integrated Optical Switch Based on Liquid Crystal Infiltration, *IEEE J. Quant. Electr.*, 41, 1197 (2005).

[13] Chigrinov, VG. *Liquid Crystal Devices: Physics and Applications*, (Artech-House, 1999).

[14] Pasechnik, SV; Chigrinov, VG; Shmeliova, DV. *Liquid Crystals, Viscous and Elastic Properties*, (Wiley-VCH, Berlin, 2009).

[15] Pozhidaev, EP. Electrooptical properties of deformed-helix ferroelectric liquid crystal display cells, *SPIE Digest*, 4511, 92 (2001).

[16] Presnyakov, Vladimir; Liu, Zhijian; Vladimir, G. Chigrinov, Fast optical retarder using deformed-helical ferroelectric liquid crystals, *Proc. SPIE*, 5970, 426 (2005).

[17] Zhang, A; Chan, KT; Demokan, MS; Chan, V; Chan, P; Kwok, HS; Chan, A. Integrated liquid crystal optical switch based on total internal reflection, *Appl. Phys. Lett.*, 86, 211108 (2005).

[18] Xu, P; Chigrinov, V; Kwok, HS. Optical analysis of a liquid-crystal switch system based on total internal reflection, *J. Opt. Soc. Am.*, 25, 866 (2008).

[19] Semenova, Yu; Panarin, Yu; Farrell, G; Dovgalets, S. Liquid crystal Based Optical Switches, *Mol. Cryst. Liq. Cryst.*, 413, 385 (2004).

[20] Zhuang, Z; Kim, Y; Patel, J. Bistable twisted nematic liquid crystal optical switch, *Appl. Phys. Lett*, 75, 3008 (1999).

[21] Muravsky, Al; Chigrinov, V. Optical Switch based on Nematic Liquid Crystals, *IDW'05 Digest*, p. 223 (2005).

[22] Maksimochkin, AG; Pasechnik, SV; Tsvetkov, VA; Yakovlev, DA; Maksimochkin, G; Chigrinov, VG. Electrically controlled switching of light beams in the plane of liquid crystal layer, *Optics Communs.*, 270, 273 (2007).

[23] Zhuang, Z; Suh, SW; Patel, JS. Polarization controller using nematic liquid crystals, *Opt. Lett.*, 24, 694 (1999).

[24] Ignacio Moreno; José, Luis Martínez; Jeffrey, A Davis. Two-dimensional polarization rotator using a twisted-nematic liquid-crystal display, *Appl. Opt.*, 46, 881 (2007).

[25] Muravsky, Alexander; Murauski; Anatoli, Chigrinov, Vladimir; Kwok; Hoi-Sing. Light Printing of Grayscale Pixel Images on Optical Rewritable Electronic Paper, *Jpn. J. Appl. Phys.*, 47, 6347 (2008).

[26] http://www.amstechnologies.com/fileadmin/amsmedia/downloads/1073_VOA-LowTDL.pdf

[27] Du, F; Lu, Y; Ren, H; et al. Polymer- Stabilized Cholesteric Liquid Crystal for Polarization –Independent Variable Optical Attenuator", *Jpn. J. Appl. Phys.*, 43, 7083 (2004).

[28] Wu, YH; Liang, X; Lu, YQ; Du, F; Lin, YH; Wu, ST. Variable optical attenuator with a polymer-stabilized dual-frequency liquid crystal, *Appl. Opt.*, 44, 4394 (2005).

[29] Hirabayashi, Katsuhiko; Amano, Chikara. Liquid-Crystal Level Equalizer Arrays on Fiber Arrays, *IEEE Phot. Techn. Letts.*, 16, 527 (2004).

[30] Haurylau, M; Anderson, S; Marshall, K; Fauchet, P. Electrical modulation of silicon-based two-dimensional photonic bandgap structures, *Appl. Phys. Lett.*, 88, 061103 (2006).

[31] Alagappan, G; Sun, X; Yu, M; Shum, P; Engelsen, D. Tunable Dispersion Properties of Liquid Crystal Infiltrated Into a Two-Dimensional Photonic Crystal, *IEEE J. Quant. Electr.*, 42, 404 (2006).

[32] Weiss, S; Ouyang, H; Zhang, J; Fauchet, P. Electrical and thermal modulation of silicon photonic bandgap microcavities containing liquid crystals, *Opt. Expr.*, 13, 1090 (2005).

[33] Ignacio Del Villar; Ignacio R. Matías; Francisco J. Arregui. Analysis of one-dimensional photonic band gap structures with a liquid crystal defect towards development of fiber-optic tunable wavelength filters, *Opt. Expr.*, 11, 430 (2003).

[34] Sireto, L; Coppola, G; Abatte, G; Righini, GC; Otón, JM. Electro-optical switch and continuously tunable filter based on a Bragg grating in a planar waveguide with liquid crystal overlayer, *Opt.Eng.*, 41, 2890 (2002).

[35] Chigrinov, VG; Kozenkov, VM; Kwok, HS. *Photoalignment of Liquid Crystalline Materials: Physics and Applications*, Wiley, (2008).
[36] Presnyakov, V; Liu, Z; Chigrinov, V. Infiltration of photonic crystal fiber with liquid crystals, *Proc. SPIE,* 6017, 102 (2005).
[37] Ertmana, Sławomir; Srivastava, Abhishek K; Chigrinov, Vladimir G; Chychłowskia, Miłosz S; Wolinski, Tomasz R. Patterned alignment of liquid crystal molecules in silica micro-capillaries, *Liq. Cryst*, 40, 1 (2013).
[38] Poon, A; Chao, Li; Ning, Ma; Lik, Lau Siu; Tong, D; Chigrinov, V. Photonics Filters, Switches and Subsystems for Next-generation Optical Networks, *HKIE Transactions*,11,60 (2004).
[39] Presnyakov, Vladimir; Asatryan, Karen; Galstian, Tigran; Chigrinov, Vladimir. Optical polarization grating induced liquid crystal micro-structure using azo-dye command layer, *Optics Express*, 14, 10558 (2006).
[40] Oh, C; Escuti, MJ. Achromatic Diffraction Using Reactive Mesogen Polarization Gratings, *SID'07 Digest*, p. 1401 (2007).
[41] Beeckman, Jeroen; Neyts, Kristiaan; Vanbrabant, Pieter JM. Liquid-crystal photonic applications, *Optical Engineering*, 50(8), 081202 (2011).
[42] Hecht, E. *Optics*, Addison Wesley, 2002.
[43] Valyukh, Sergiy; Chigrinov, Vladimir; Kwok, Hoi Sing. A Liquid Crystal Lens with Non-uniform Anchoring Energy, *SID'08 Digest*, p. 659, 2008.
[44] Ren, Hongwen; Fan, Yun-Hsing; Gauza, Sebastian; Wu, Shin-Tson. Tunable-focus flat liquid crystal spherical lens, *Appl. Phys. Lett.*, 84, 4789 (2004).
[45] Presnyakov, VV; Asatryan, KE; Galstian, T; Tork, A. Polymer-stabilized liquid crystal for tunable microlens applications, *Opt. Express*, 10, 865, 2002.
[46] Ye, M; Wang, B; Sato, S. Liquid crystal lens with a focal length that is variable in a wide range, *Appl. Opt.*, 43, 6407, 2004.
[47] Tseng, Man-Chun; Fan; Fan, Lee, Chung-Yung; Murauski, Anatoli; Chigrinov, Vladimir; Kwok, Hoi-Sing. Tunable lens by spatially varying liquid crystal pretilt angles, *J. Appl. Phys.*, 109, pp. 083109-1 (2011).
[48] Fan, Fan; Srivastava, Abhishek Kumar; Du, Tao; Tseng, Man Chun; Chigrinov, Vladimir; Kwok, Hoi Sing. Low voltage tunable liquid crystal lens, *Opt. Letts.*, 38, 4116 (2013).
[49] Wang, Xiao-Qian; Srivastava, Abhishek K; Chigrinov, Vladimir G; Kwok, Hoi-Sing. Switchable Fresnel lens based on micropatterned alignment, *Opt. Letts.*, 38, 1775 (2013).
[50] Sergiy, Valyukh; Vladimir Grigorievich, Chigrinov; Oleksandr Siobodyanyuk, Kyiv. Liquid Crystal Display, US Provisional Patent Application No. 61 /853,734 filed on 11 April 2013.
[51] Beeckman, Jeroen; Neyts, Kristiaan; Vanbrabant, Pieter JM. Liquid-crystal photonic applications, *Opt. Eng.*, 50(8), 081202 (2011).
[52] Huang, Yi-Pai; Chen, Chih-Wei; Huang, Yi-Ching. Superzone Fresnel Liquid Crystal Lens for Temporal Scanning Auto-Stereoscopic Display, *Journ. of Displ. Techn.*, 8, 650(2012).
[53] Chen, Chih-Wei; Cho; Myungjin; Huang, Yi-Pai; Javidi, B. Three-dimensional imaging with axially distributed sensing using electronically controlled liquid crystal lens, *Opt. Letts*, 37, 4125 (2012).

[54] Ye, Mao; Sato, Susumu. Liquid Crystal Lens with Insulator Layers for Focusing Light Waves of Arbitrary Polarizations, *Jpn. J. Appl. Phys.*,42, 6439 (2003).

[55] Wang, Bin; Ye, Mao; Homma, Michinori; Nose, Toshiaki; Sato, Susumu. Liquid Crystal Lens with Spherical Electrode, *Jpn. J. Appl. Phys.*, 41, L 1232 (2002).

[56] Kim, Jae-Hoon; Kumar, Satyendra. Fabrication of Electrically Controllable Microlens, Array Using Liquid Crystals, *L. Lightwave Techn.*, 23, 628 (2005).

[57] Ren, Hongwen; Fox, David W; Wu, Benjamin; Wu, Shin-Tson. Liquid crystal lens with large focal length tunability and low operating voltage, *Opt. Expr.*, 15, 11328 (2007).

[58] Presnyakov, Vladimir V; Galstian, Tigran V. Electrically tunable polymer stabilized liquid-crystal lens, *J. Appl. Phys.*, 97, 103101 (2005).

[59] Fraden, J. *Handbook of Modern Sensor: Physics, Designs and Applications*, Springer New York (2004).

[60] Biga, FY; Luks, FI; et al., Medical Displays, 81-149 in book "*Liquid Crystals: Frontiers in Biomedical Applications*" Ed. S. J. Woltman, G. R Crawford, G.D.Jay, World Scientific Publishing, Singapore, (2007).

[61] Brodzeli, Zourab; Silvestri, Leonardo; Michie, Andrew; Guo, Qi; Pozhidaev, Evgeny P; Chigrinov, Vladimir; Ladouceur, François. Sensors at Your Fibre Tips: A Novel Liquid Crystal-Based Photonic Transducer for Sensing Systems, *J. Lighwave Techn.*, 31, 2940 (2013).

[62] Zhao, Xiaojin; Boussaid, Farid; Bermak, Amine; Vladimir G. Chigrinov. Thin Photo-Patterned Micropolarizer Array for CMOS Image Sensors, *IEEE Phot. Techn. Lett.*, 21, 805 (2009).

[63] Zhao, Xiaojin ; Bermak, Amine; Boussaid, Farid; Du, Tao; Chigrinov, Vladimir G. High-resolution photoaligned liquid-crystal micropolarizer array for polarization imaging in visible spectrum, *Opt. Letts.*, 34, 3619 (2009).

[64] Thorn, R; et al., *Differential Pressure Flowmeters*, part 9.1.in book *"Mechanical variables measurement: Solid, Fluid, and Thermal", ed. by J.G.Webster,* CRC Press LLC, Boca Raton (2000).

[65] Woli´nski, Tomasz R; Jarmolik, Aleksandra; Bock, Wojtek J. Development of Fiber Optic Liquid Crystal Sensor for Pressure Measurement, *IEEE Trans. On Instr. and Meas.,* 48, 2 (1999)

[66] Fan, Fan; Du, Tao; Srivastava, Abhishek Kumar; Lu, Wang; Chigrinov, Vladimir; Kwok, Hoi Sing. Axially symmetric polarization converter made of patterned liquid crystal quarter wave plate, *Optics Express,* 20, 23036 (2012).

[67] Slussarenko, Sergei; Murauski, Anatoli; Du, Tao; Chigrinov, Vladimir; Marrucci, Lorenzo; Santamato, Enrico. Tunable liquid crystal q-plates with arbitrary topological charge, *Optics Express,* 19, 4085 (2011).

[68] Slussarenko, Sergei; Piccirillo, Bruno; Chigrinov, Vladimir; Marrucci, Lorenzo; Santamato, Enrico. Liquid crystal spatial-mode converters for the orbital angular momentum of light, *J. Opt.,* 15, 025406-1 (2013).

[69] Wu, Hao; Hu, Wei; Hu, Hua-chao; Lin, Xiao-wen; Zhu, Ge; Choi, Jae-Won; Chigrinov, Vladimir; Lu, Yan-qing. Arbitrary photo-patterning in liquid crystal alignments using DMD based lithography system, *Optics Express*, 20, 16684 (2012).

[70] Vladimir G. Chigrinov. Fast Switchable Liquid Crystal Cells for Field Sequential Color and 3D Displays, *OLC 2011*, Invited I14, Erevan, Armenia, September, 2011.

[71] Uchida, T; Miyashita, T. Opticaly Compensated Bend Mode with Wide Viewing Angle and Fast Response for AM-LCD, *IDW'95 Digest*, p.39 (1995).

[72] Fion, Y; Kwok, HS. Fast-response no-bias-bend liquid crystal displays using nanostructured surfaces, *Appl. Phys Letts*, 88, 063505 (2006).

[73] Li, YW; Tan, L; Kwok, HS. Field-Sequential-Color LCDs Based on Transient Modes, *SID 2008 Digest*, p. 32 (2008).

[74] Xu, Peizhi; Chigrinov, Vladimir; Kwok, Hoi Sing. Optical analysis of a liquid-crystal switch system based on total internal reflection, *JOSA A*, 25, 866 (2008).

[75] Castles, F; Morris, SM; Coles, HJ. Flexoelectro-optic properties of chiral nematic liquid crystals in the uniform standing helix configuration, *Physical Review E*, 80, 031709 (2009).

[76] Rao, L; Yan, J; Wu, ST; Yamamoto, S; Haseba, Y. Polymer-stabilized blue phase liquid crystal with a negative Kerr constant, *Appl. Phys. Lett.*, 98, 081109 (2010).

Chapter 5

CONCLUSION

Liquid crystal devices have come of age, having fulfilled their promise of several decades ago by increasingly dominating the market for displays. The new basic research in LC are based on photoalignment and fast ferroelectric electrooptic LC modes as well as bi and multi-stable LC states, which is an origin of low power memory display and photonics devices, such as E-paper. In the field of liquid crystals (LC), the main point is no longer display research, as new photonic applications of LC are emerging in telecommunication, fiber optical communication systems, sensors, switchable lenses, LC light converters and other LC photonics devices. New passive elements for fiber optical communication systems include also hybrid photonic crystal/ liquid crystal (PC/LC) structures. Such elements have begun to appear in the photonics market. Passive elements for fiber optical communication systems (DWDM components) based on LC such as switches, polarization rotators, variable optical attenuators can successfully compete with micro-electromechanical systems (MEMS), thermo-optical, opto-mechanical or acousto-optical devices. Nematic and ferroelectric LC are used in high-speed communication systems, producing elements that are extremely fast, stable, durable, of low loss, operable over a wide temperature range, and that require small operating voltages and extremely low power consumption. Good robustness of LC photonics components due to the absence of moving parts and compatibility with VLSI technology, excellent parameters in a large photonic wavelength range, whereas the complexity of the design and the cost of the device are equivalent to regular passive matrix LC displays, makes LC fiber optical devices very attractive for mass production. The book provided a state-of-the-art overview of the area, highlighting the latest advances and possible future directions. The research activities, presentations and discussions of new results in the photonic area of LC was also the main aim of the book. The book was concentrating on interesting or novel applications, or enhancing LC material properties for photonic applications. The content of the book covers both the main LC basic properties, electrooptical effects and most important LC photonics applications based on the new advanced LC technologies.

INDEX

#

A

B

C

D

E

F

G

H

I

J

K

L

M

Q

R

S

T

U

V